Contents

Introduction

The approach taken to presenting technical information to students has to move forward, especially in this high-tech computer and digital age.

Construction is an ideal medium for illustrating construction practice in a digital format. An understanding of work sequences such as the formation of a pile cap can be more simply illustrated by a series of photographs, line drawings and short text notes.

The aim of this book is to present construction practice in this simple but effective format. Current construction textbook authors are including a great number of photo images in textbook revisions and updates.

Construction Practice is totally presented in a photo image format. The majority of the construction sequences have been photographed by me and carefully selected for inclusion with line diagrams and text. A little repetition in certain chapters is inevitable as many of the construction processes overlap.

Data for the book has been collected during a building resession– the like of which has not been experienced before, either in the UK orin mainland Europe.

Sites have proved difficult to find for case studies and comparative examples in the various chapters. In the north-west of England it has been impossible to find a timber-framed house under construction, especially in the speculative housing market. Building projects have simply ground to a halt.

Luckily I have maintained good contacts with the larger contractors involved in large commercial projects, school building programmes and small one-off projects.

I have travelled abroad extensively to photograph projects in Holland, France, Italy and Spain either on holiday or on specific site visits–my wife has also become a construction expert in trying to stop me visiting sites whilst on holiday. It has cost me a fortune in shopping expenses whilst I visit a nearby building site for two hours or more.

A number of large north-west projects feature in the text. This includes the Liverpool One Project, Mann Island and major commercial projects in Manchester, Derby and Sheffield.

The presentation of a textbook on construction technology is a new venture since my previous books have been on management aspects. Management of the construction process is an integral part of understanding work sequences–especially for building technicians and graduates alike.

I hope that students, practitioners and professionals serving the construction industry accept this different approach to construction practice.

A series of project case studies have been included which have attempted to cover a range of construction operations from foundation to completion. It is hoped that further case studies may be included in future revisions and editions of this book.

Hope you like it.

Brian Cooke

Acknowledgements

The author would like to express a special thanks to Paul Hodgkinson for his expertise in producing the page layouts and diagrams. Without his dedication the book would not have been possible.

I am very grateful to the production team at Wiley-Blackwell with special thanks to Paul Beverley for the proof corrections. My final thanks go to Dr Paul Sayer for having faith in me to produce a first book on *Construction Practice*.

The book would not have been possible without access to projects being constructed by BAM. Bullivant, Carillion, Countryside Properties, Laing-O'Rourke, McAlpine Projects, McVeigh Insulations, Premier Waste, Taylor-Wimpey, Wates and others.

Offer to Universities/Technical Institutions/Educational establishments.

The author is willing to consider written invitations to any educational establishments throughout the United Kingdom, Europe, the Far East and China. He is willing to give lectures on aspects of construction management, project planning and the building technology areas outlined in *Construction Practice*. Written requests to be made to the publisher – Wiley-Blackwell, care of the Publisher for construction.

November 2010

CHAPTER ONE

ESTABLISHING THE SITE

Construction Practice. Brian Cooke

1.0 Overview

This chapter deals with the management of a construction project at the pre-contract stage of a project. It is at the pre-contract stage that decisions are made in relation to site layout planning, construction methods and safety procedures to be implemented. The above areas are outlined in relation to a range of site situations.

Site layout planning considerations include the factors to be considered when creating a site layout plan. The importance of crane locations is outlined in relation to a major project in Liverpool.

Site signage is highlighted together with the cost of providing signs as part of the contract preliminaries.

A series of site accommodation situations are illustrated in relation to six projects. Two site layout plans are indicated for both a small and a medium-sized project. Good site practices are highlighted.

The proceedures to be followed when visiting a construction project are outlined and the importance of the induction process is covered, together with a summary of the site rules.

An understanding of the build-up of contract preliminaries is essential in order to relate the allowances in the preliminaries bill to the site situation.

An explanation of time-and fixed-related preliminaries is given together with the build-up of site accommodation costs for a contract at the tender stage.

Site logistical problems in relation to an inner city development are discussed in a practical site situation. Decisions in relation to site access and surrounding buildings are discussed.

It is considered that site managers must be familiar with allowances in the contract bills for site facilities and services. Far too often the site manager is blamed for overspending on site preliminaries when all the time there were insufficient monies available.

1.1 Procedure prior to commencing a project

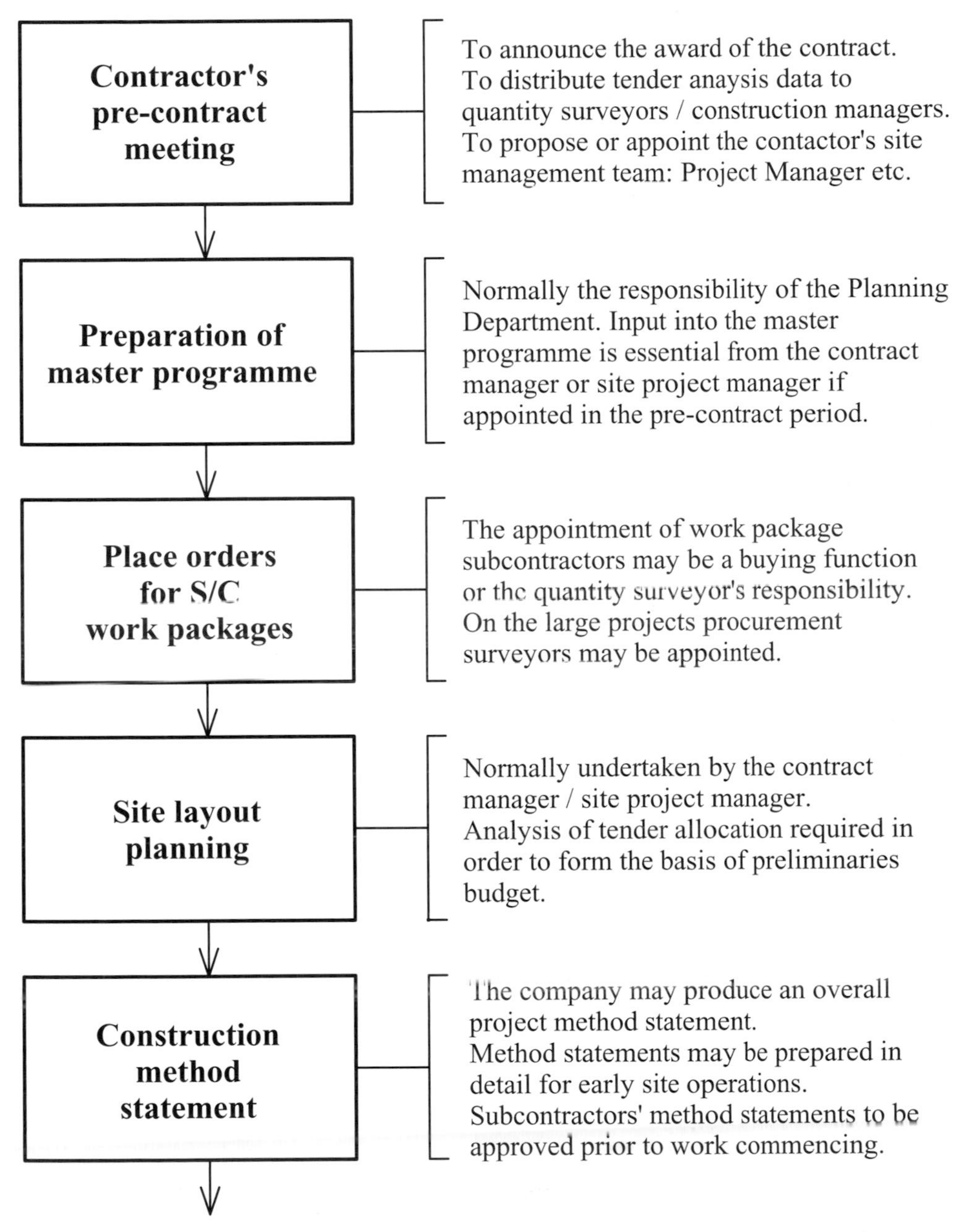

Depending on the size of the organisation, the above procedures may occur in a different sequence or be combined activities.

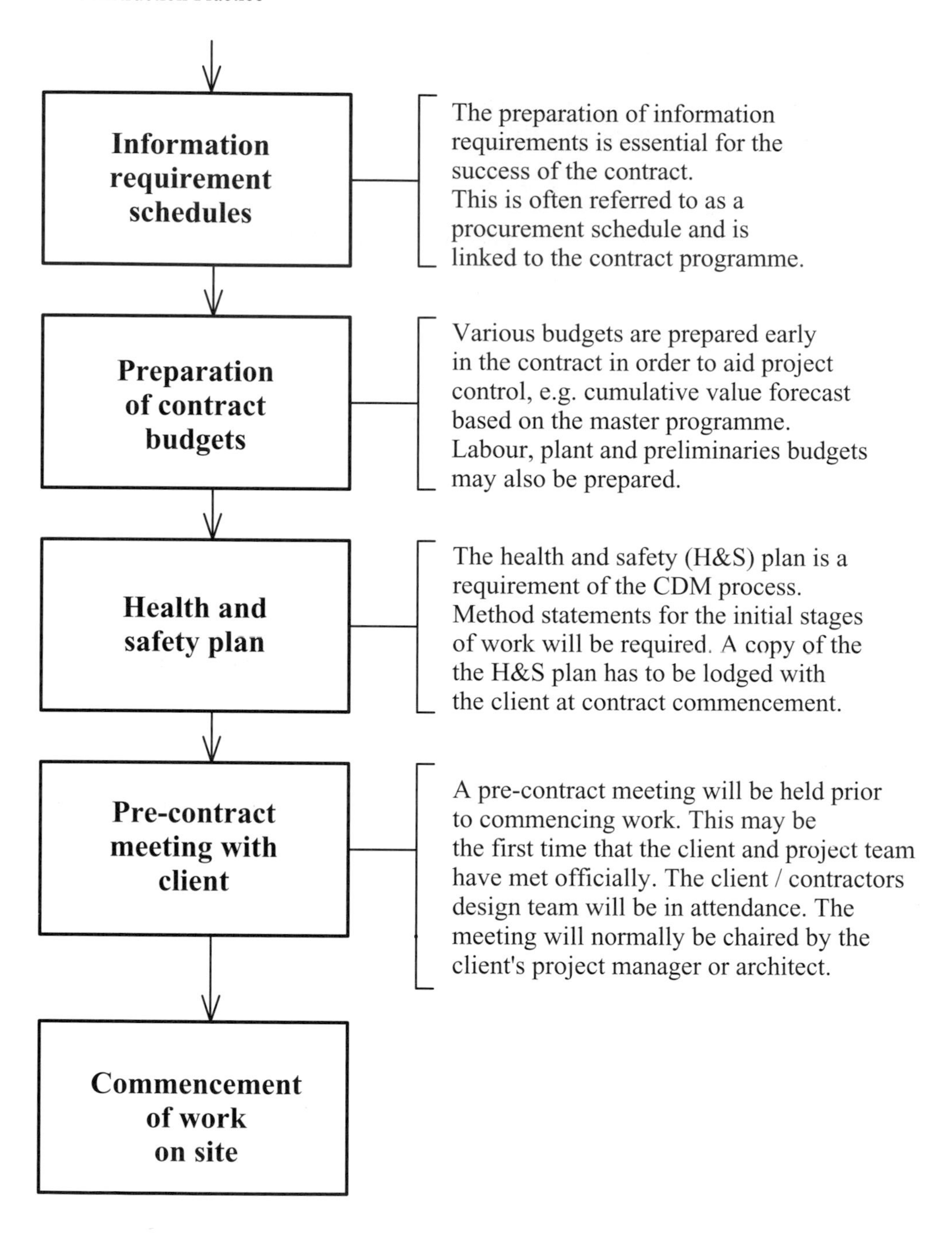
Information requirement schedules
The preparation of information requirements is essential for the success of the contract.
This is often referred to as a procurement schedule and is linked to the contract programme.
Preparation of contract budgets
Various budgets are prepared early in the contract in order to aid project control, e.g. cumulative value forecast based on the master programme.
Labour, plant and preliminaries budgets may also be prepared.
Health and safety plan
The health and safety (H&S) plan is a requirement of the CDM process.
Method statements for the initial stages of work will be required. A copy of the the H&S plan has to be lodged with the client at contract commencement.
Pre-contract meeting with client
A pre-contract meeting will be held prior to commencing work. This may be the first time that the client and project team have met officially. The client / contractors design team will be in attendance. The meeting will normally be chaired by the client's project manager or architect.
Commencement of work on site

1.2 Site layout planning considerations

There is a direct link between site layout planning and materials management, and this can affect the degree of loss and wastage created on site.

Site layout planning starts at the tender stage of a project. Decisions are made in the preliminaries estimate build-up for allowances to be included in the tender for materials management and site layout proposals. At the tender adjudication stage these allowances may often be reduced in order to make the bid more competitive. Getting the balance right is difficult to achieve in practice.

Once the contract has been awarded, the contractor has to ensure that he keeps within the preliminaries expenditure budget during the contract. Monetary losses on contract preliminaries are a common occurrence in both the medium and large contracting organisations. A preliminaries budget should be prepared prior to commencing the project. The actual preliminaries monetary release should be matched with the preliminaries expenditure on a monthly and cost-to-ate basis.

Considerations at the contract commencement stage

The contractor should consider the following points:

- What allowances have finally been included in the priced bills for preliminary expenditure?
- An analysis of the proposed preliminaries allowances should be provided from the estimating and surveying section.
- What statutory requirements must be provided in respect of the Health and Safety at Work Act–in relation to welfare and site facilities, offices and accommodation to be provided?
- A site layout plan, traffic movement plan and fire plan may be prepared at this stage.
- A site hazard board should be located adjacent to the site mess and office facilities

A site layout plan should be prepared in order to show the contractor's proposals. The main input into this task is normally undertaken by the contractor's project manager or site manager if appointed prior to the commencement of site work.

The site layout proposal may need to be approved by the client or his project manager prior to the commencement of work.

The site layout plan may indicate:

- location of offices and accommodation. This includes consideration of the main contractor's accommodation, provision of mess facilities for operatives and subcontractors; requirements for welfare facilities including drying rooms, toilet blocks and signing in facilities. Provision may also be provided for the induction process. Lockable storage containers may also be considered for more valuable materials.
- proposals for site access and egress for delivery vehicles and site personnel, providing directory signs to aid deliveries.
- position of material storage areas adjacent to the works and areas for the disposal of waste materials. Separate areas may be allocated for specific subcontractors' materials e.g. steelwork unloading areas prior to the fixing of the steelwork etc. The need for material storage areas will change during the project–as one operation finishes another will start and more materials will be delivered for the next stage of the works.
- vehicle parking areas for site staff and operatives' vehicles (if site space is available on site). Vehicles parked in streets adjacent to a building site often result in disputes with local residents. The contractor will be responsible for damage to footpaths and verges adjacent to the site.
- pedestrian access routes to the work areas. These should be clearly defined and marked on the item layout plan. These should be pointed out to operatives at the induction stage. Security entry and exit turnstiles may be provided complete with hand recognition equipment.
- space around the building for the provision of scaffolding, hoists and access areas to the base of hoists. Loading platforms require additional space in front of them for loading facilities and plant movement.
- site signage, i.e. directory signs, safety signs and warning signs. A range of signs in current use on construction sites is illustrated in this chapter. The cost associated with the provision of signs should be considered carefully at the preliminaries pricing stage.
- security fencing and site hoardings including secure entry gates. The use of 24-hour video cameras may be considered as part of the security arrangements.
- site mixing areas and mortar silos.

A separate copy of the site layout plan should indicate the provisions for fire safety and access for firefighting equipment. Site operatives should be made aware of the procedures in case of fire at the site induction stage.

1.3 Location of tower cranes

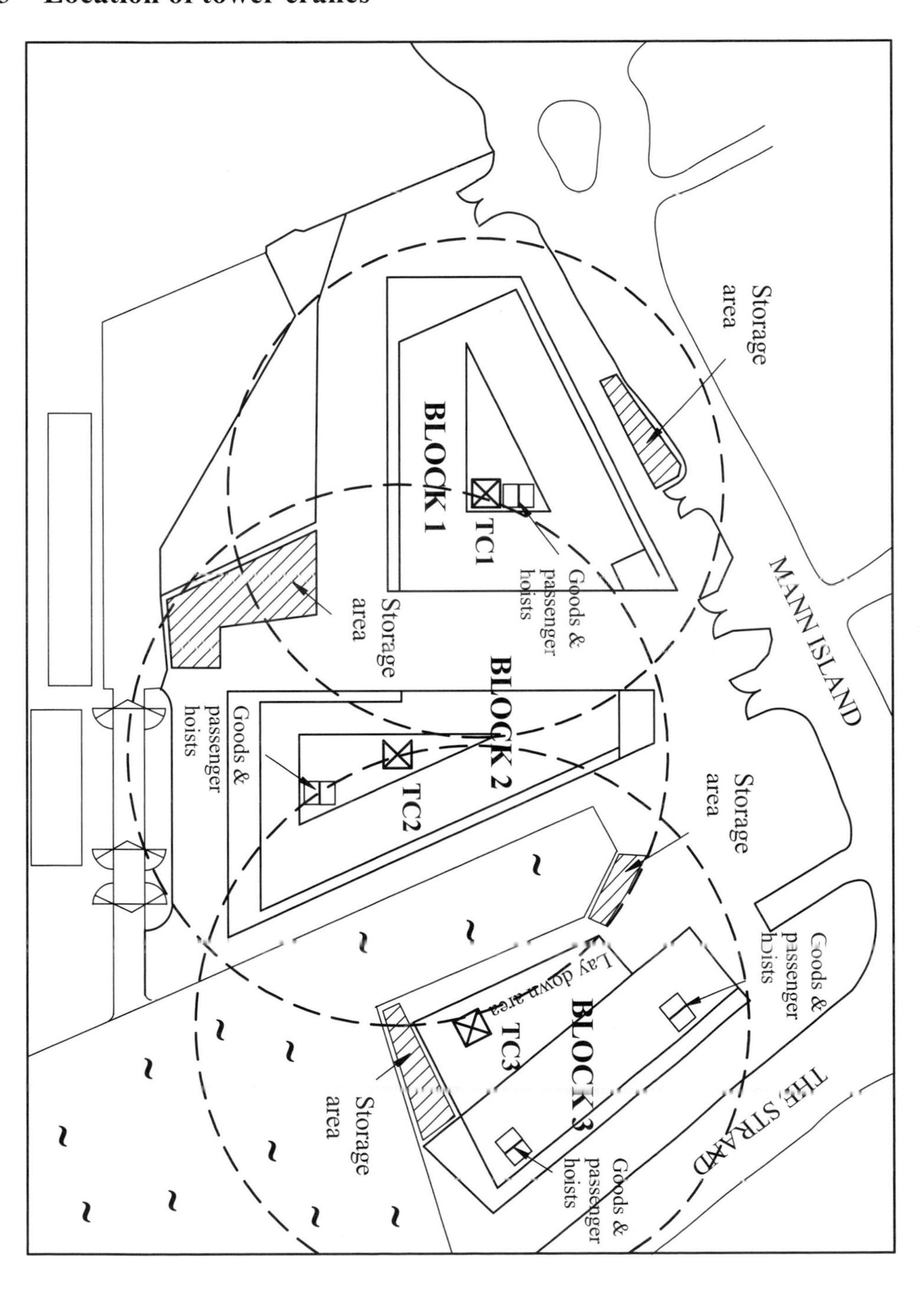

View of tower cranes

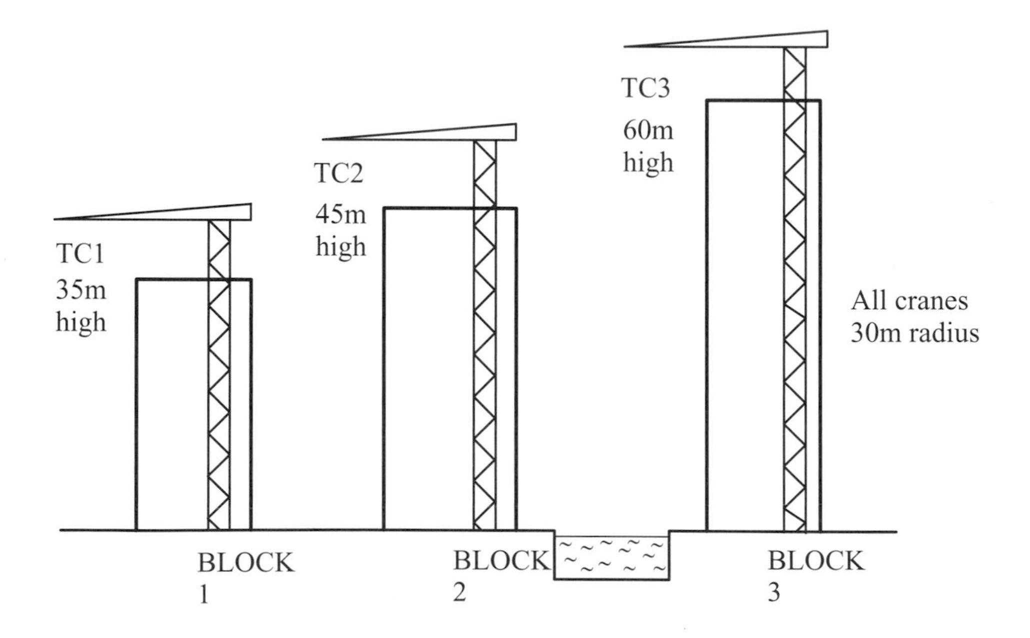

The siting of major plant items such as tower cranes may be shown on the layout plan. The location of the tower crane is critical to site access roads and unloading areas within the radius of the crane. Four main storage areas have been identified on the layout plan illustrated.

On the above £100 million project each of the tower cranes serves an individual block. The single and twin-core staircase towers are each 15 storeys in height (Block 3 on the plan). Concreting operations were undertaken by tower crane and skips.

Lay down storage areas have been indicated together with the location of passenger and goods hoist locations.

1.4 Site signage examples

DIRECTORY AND INFORMATION SIGNS

Allowance for provision of site signage costs

The allowance for site establishment, offices, site accomodation and items such as site security are allowed for in the contract preliminaries. Allowance must also be included for site hoardings and protection. Monies must also be included for the provision of site directory, information and safety signage. This latter provision is not a cheap item. Signs have to be purchased, fixed in position and possibly written off during the project as many may be site specific and not reusable.

SAFETY SIGNS

Information signs

Provision of site signage

The preliminaries allowance for providing 15 site signs may be built up of :

Cost of providing 15 signs at £50 average	=	£ 750
Labour fixing & removal of signs–15 hrs at £30	=	£ 450
Total prelims allowance for signage	=	£ 1200

Personnel hygiene facilities

1.5 Site accommodation situations

Situation one

Site office facilities located in a basement area–after the pouring of the basement slab and erection of the tower crane. The roof of the office has been used for storing fabric reinforcement. The new building and basement are located in an Amsterdam street between existing four-storey buildings. Access at the front of the building is directly onto the footpath and street.

Situation two

Minimal site accommodation on a £650,000 project. Site offices consist of a site manager's office come meeting room. Adjacent is a further cabin for the subcontract labour, toilet block and lockable storage container. The photograph shows the piling, subcontractors accommodation, office and mess.

Situation three

Nine stackable office units covering three floors with staircase access. Project involves the construction of an £8m four-storey office block with limited site space. Offices located on adjacent land at rear of project. Car parking facilities for staff are provided on an adjacent area.

Situation four

Site accommodation requirements for a £6m residential housing project. Two two-tier offices with central access stairways. Accommodation consists of site managers' office come meeting room, toilets, operatives' canteen and drying room. Four lockable containers are provided within the compound area.

Situation five

Seven mobile office units on a £20m new college project. Four further similar units on site plus security office and induction room. For the project size and value the number of offices provided appears to be a somewhat expensive solution.

Situation six

Basic site accommodation, toilet and store all in one lightweight mobile unit.

Excellent small mobile unit from the Cabin Centre.

1.6 Site layout planning examples

A more simplistic approach is shown below for the site layout proposals for a new church project on a confined site. The project involved the construction of a timber portal framed building, access road and car park area.

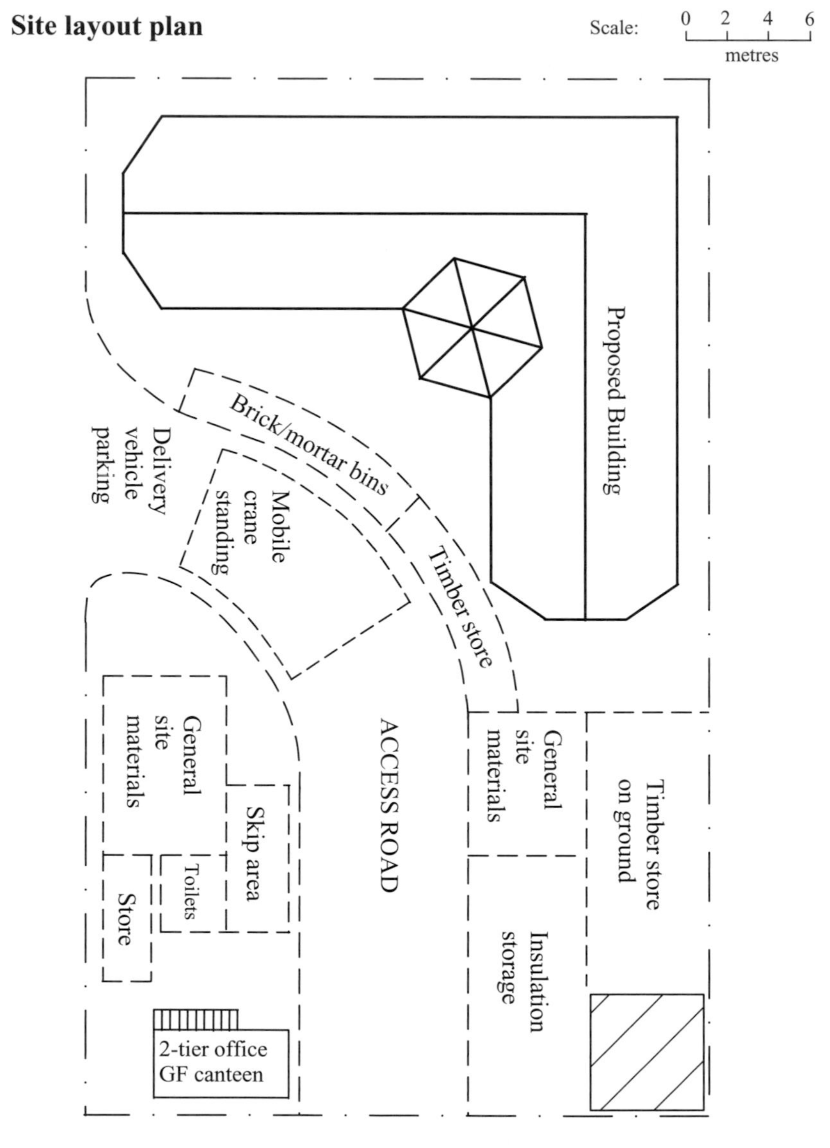

Disley church project

The proposed site layout plan opposite is for a £14m college project is illustrated. A separate fire plan for the project was also proposed by the contractor.

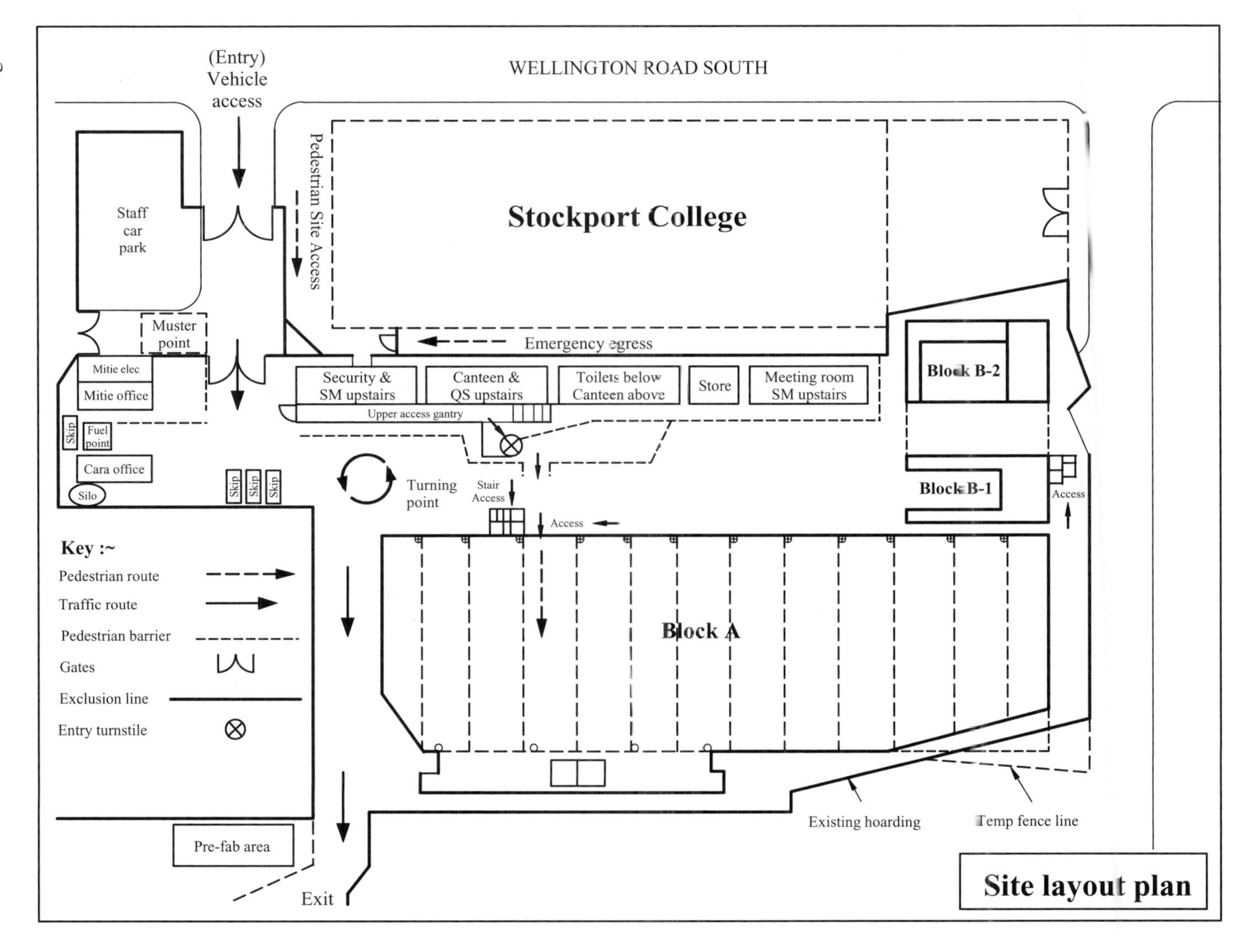

(Entry) Vehicle access
WELLINGTON ROAD SOUTH
Staff car park
Pedestrian Site Access
Stockport College
Muster point
Emergency egress
Mitie elec
Mitie office
Security & SM upstairs
Canteen & QS upstairs
Toilets below Canteen above
Store
Meeting room SM upstairs
Block B-2
Upper access gantry
Skip
Fuel point
Cara office
Silo
Skip
Skip
Skip
Turning point
Stair Access
Access
Block B-1
Access
Key :~
Pedestrian route
Traffic route
Pedestrian barrier
Gates
Exclusion line
Entry turnstile
Block A
Existing hoarding
Temp fence line
Pre-fab area
Exit
Site layout plan

Site office layout on Stockport College project (£14m project)

A 'mega' block of site accommodation on a large Liverpool project–sixty mobile offices for client and contractor

As a project moves forward, further operations come on stream. Materials for cladding and glazing come on site.

Site storage area for storey height granite external cladding panels

Canti boxes being loaded for positioning in the building

Good practice layout plan

1.7 Procedure when visiting a project

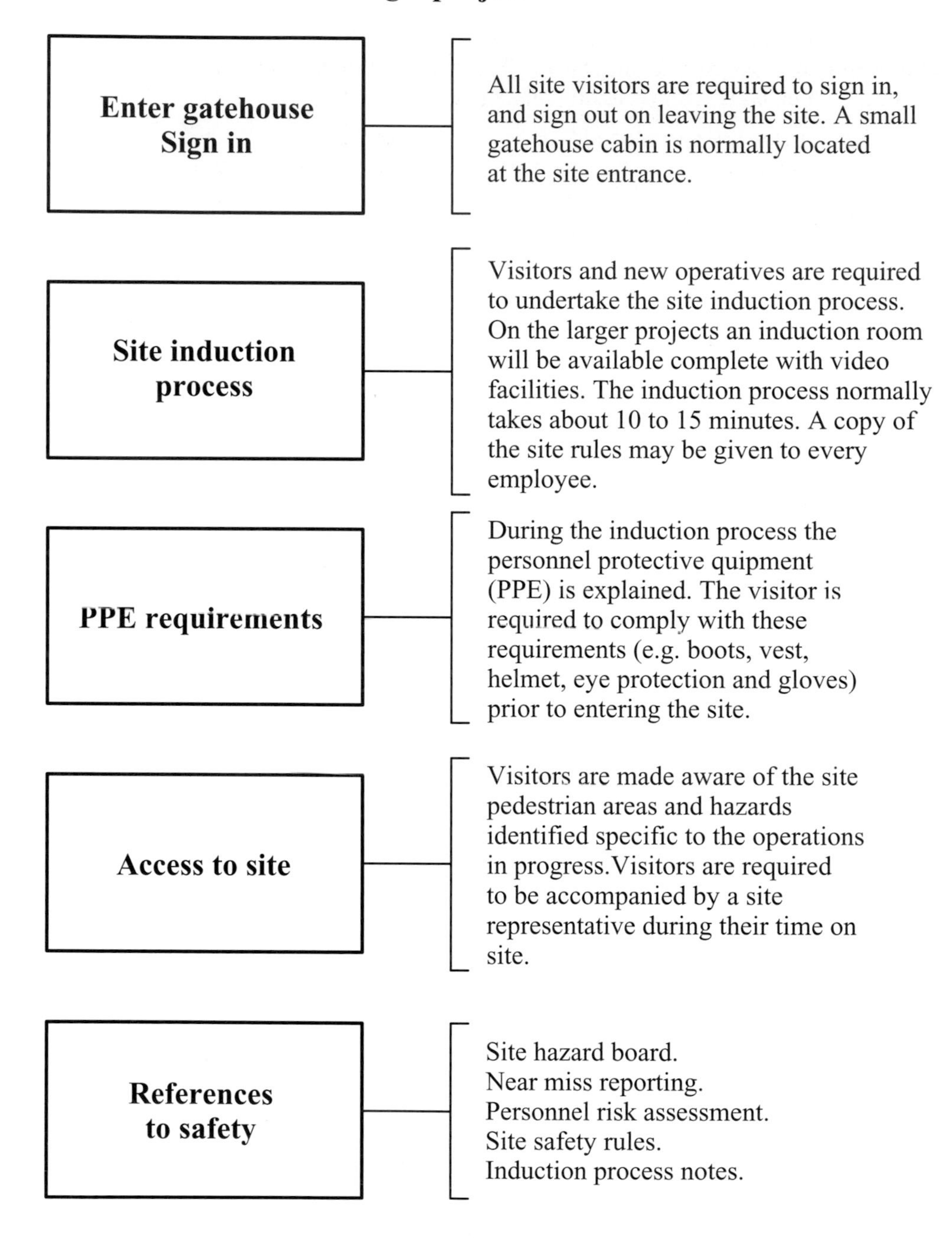

1.8 Site induction procedures

The responsibility for the site induction process is undertaken by a member of the contractor's site management team. This may fall on the site safety representative or be allocated to a construction manager.

On notifiable projects (lasting more than 30 days) it is the principal contractor's statutory duty to ensure that every construction worker on site is provided with a suitable site induction. (EDM 2007–regulation 22/2)

The purpose of the induction process is:

- to indicate the site layout and welfare facilities
- to highlight emergency site procedures
- to identify site hazards and direct attention to the daily hazard boards
- to outline the importance of the site rules as applicable to the project, a copy of which may be given
- to indicate the importance of subcontractors' and main contractor's method statements
- to outline the contractor's safety policy
- to complete the necessary registration procedures for the participant (i.e. competence levels, CSC cards etc.).

The site fire plan will be outlined in order to identify escape routes, muster points and the location of firefighting facilities.

The following is an extract from a set of site rules.

- No person permitted on site unless he has been inducted.
- PPE (personnal protective equipment) shall be worn.
- Any person interfering with or misusing fixtures, fittings, scaffold or equipment will be excluded from site.
- Safety signs and notices must be complied with.
- Radios, stereos and iPods are not permitted to be used on site.
- All site personnel are required to comply with their employer's method statements and risk assessments.
- The consumption of alcohol or drugs is prohibited.
- No person to operate any mechanical plant or equipment unless they have been fully trained.
- The wearing of shorts is not permitted.
- Smoking may only take place in designated areas.
- Food is only to be consumed in the designated mess area.
- Every accident or near miss occurence must be reported.
- Site fire and emergency alarms and equipment must be followed.

Provisions of on-site facilities

Security office at site entrance

Signing in on entry to site

Induction room with video

Safety board in office reception

Hand recognition check system

Gate entry to site

1.9 Contract preliminaries Terminology

The contract preliminaries section of the bill allows the contractor to enter a sum of money against a preliminary bill item for site accommodation. A single sum of money is entered at the tender stage.

Site accommodation (and site establishment costs) are priced by the estimator on a ***fixed-related cost*** build up plus a ***time-related cost.*** The total makes up the single entry in the bill.

The fixed-related costs are the costs:

- setting up the site office compound area
- preparing, hardcoring or tarmacing the area
- erecting any fencing around the area
- laying on site services–water/drainage/power
- delivery and erection of offices, mess huts, toilets, stores, security cabin etc.
- erection and fit out of the above
- dismantling/removal from site and reclaiming the area.

The time-related costs relate to the number of weeks that the office/equipment are on site with respect to:

- the hire costs of all accommodation
- the weekly servicing costs for the same, i.e. costs of running and maintaining the office services.

The estimator will liaise with the project manager/commercial manager regarding site office and compound requirements. Each contractor preparing a bid will take a different approach to the assessment of staffing levels and site accommodation requirements.

Decisions will have to be made relating to the number of cabins and type of accommodation to be provided. The estimator will be sufficiently experienced to assess site requirements based on the pricing of similar projects. Advice may be available from a contracts manager or the commercial manager in charge of the estimate.

Over-assessment of contract preliminaries will result in the final tender submission being less competitive. Getting the balance right is difficult.

Build-up of site preliminaries costs

At tender adjudication stage, final tender adjustments are the responsibility of senior management. Normally the total cost or value of all contract preliminaries may represent 5–15 per cent of the contract sum. The contractor may be required to submit a preliminaries breakdown or analysis on commencement of the contract. This is to aid the valuation process as fixed cost preliminary items, for setting up the site, will be claimed in the early project valuations.

Pricing for site accommodation–practical case

Project scenario

A contractor is preparing a bid for a £8m office block on a competitive tender.

Practical case

Contract period 40 weeks. The form of contract is to be design and build (client-led design). It is envisaged that the contractor will undertake brickwork and joinery items with his own labour. The remaining work will be undertaken as a series of work packages.

Task

Prepare an assessment of the preliminaries build-up for site establishment tender stage.

Assumptions

Adequate space on site for accommodation based on tiered offices as illustrated opposite.

The proposals for the use of tiered accommodation consisting of six office / cabins with a central staircase access is shown.

A build-up of the fixed costs of establishing the site and the time-related costs for the hire of the units throughout the contract period of 40 weeks is shown.

The staffing requirements for the project are assumed to be:

- project manager–40 weeks
- assistant project manager–30 weeks
- site manager–40 weeks
- site engineer–18 weeks
- site-based quantity surveyor–40 weeks (part visiting QS role and planner)
- visiting safety management/senior management – head office overhead.

Site accommodation layout proposed

Units B, C, D & E – 9m x 3m

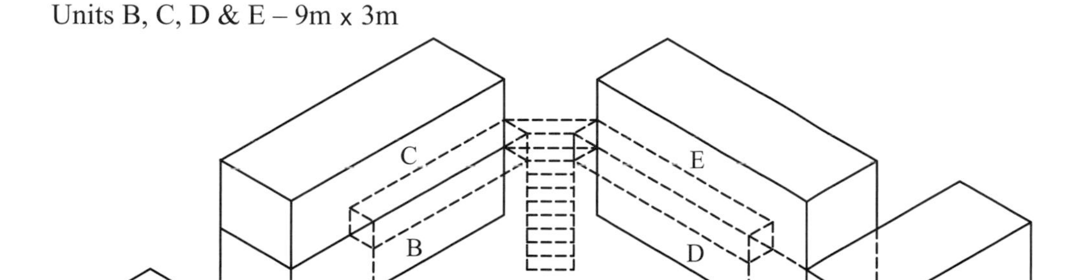

A – site security/signing in cabin

B – operatives' mess - including drying room

C – project manager/assistant project manager

D – site-based supervisory staff/engineer/visiting QS

E – Meeting room

F – toilet block – male/female

Area containing site offices to be formed with timber hoarding/gates. Central area for accommodation of four staff vehicles. All services to compound to be provided.

Hire rate to be based on six units

4 No. 9m x 3m mobile office units

1 No. 6m x 3m double toilet block

1 No. 3m x 3m site security cabin

Hire rates for various types of site accommodation based on late 2009 quotation are as follows:

9m x 3m office	£40/week
9m x 3m canteen/drying room	£40/week
9m x 3m toilet block/sinks	£60/week
9m x 3m store	£20/week
Staircase and balcony platform	£10/week
Delivery including stacking two- two-stacked offices	£500
Offices included standard furniture delivery and installation per unit	£150/unit

Final build-up of costs for inclusion in tender

Fixed cost areas			
• Prepare area for compound/stone up/tarmac	Item		2500
• Perimeter hoarding and gates	Item 80m	£80	6400
• Services–drainage/electricity/water/telephones	Item		1600
• Delivery and erection of tiered accommodation (fully fitted out offices assumed)	Item 5 units	£300	1500
• Dismantle offices/remove hoardings–remove from site	Item		1000
• Reclaim compound area	Item		1000
Total fixed costs £14000		£	14000
Time-related costs / 40 weeks	Cost/ week		
• 4 No. 9m x 3m mobile units 4 No. at £60/w	£160	x 40	6400
• 1 No. 9m x 3m toilet block 1 No. at £60/w	£60	x 40	2400
Maintenance costs	£50	x 40	2000
• 1 No. 3m x 3m security block 1 No. at £20/w	£20	x 40	800
• Weekly running costs of services/electricity/cleaning costs/week	£400	x 40	1600
Contingency			1000
		£	28600

Total fixed costs = £14000
Total time related = £28600

Total time & fixed costs = £42600

Example of site accommodation availability

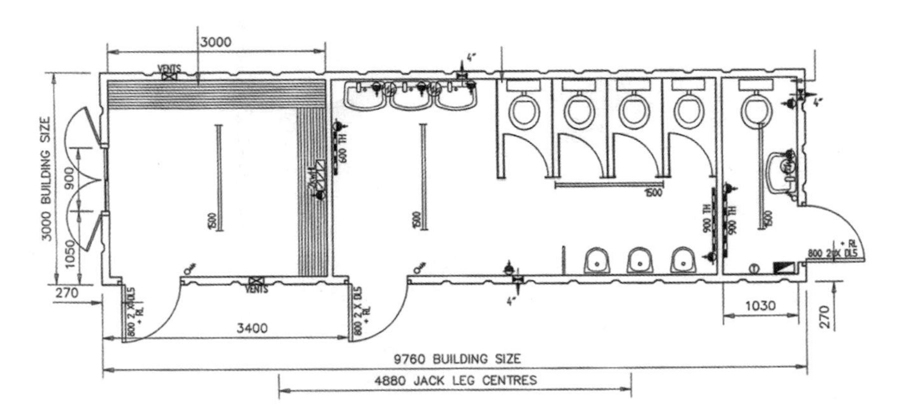

Toilets and drying room

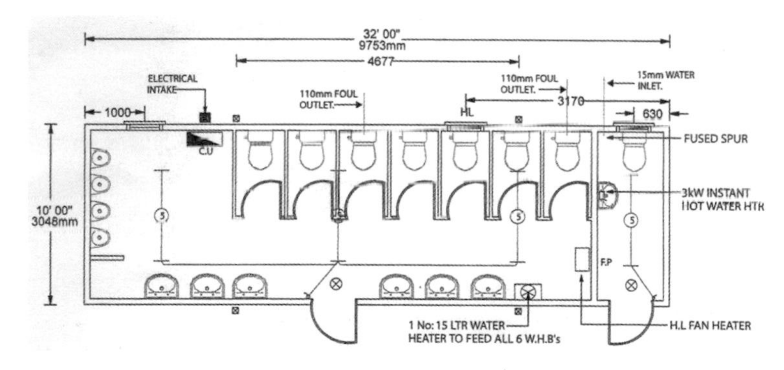

Toilets

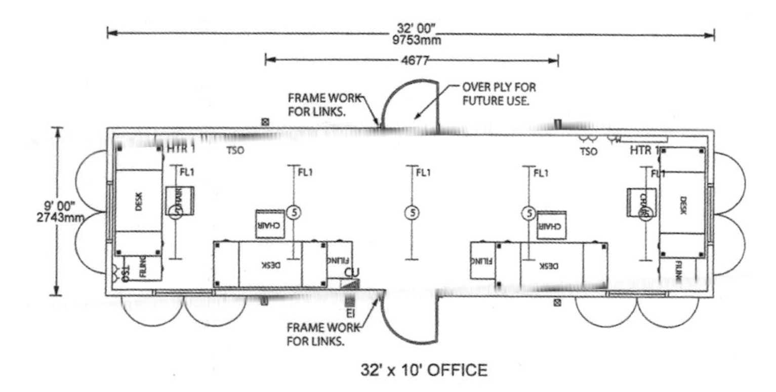

Office block

1.10 Site logistics
Ten-storey inner city office development

Description of project

The project involves the construction of a 10-storey inner city office block in the heart of the Manchester business centre.

The £8m building is of steel-framed construction incorporating metal deck concreted floors. Externally the building is clad with a glass curtain wall consisting of storey-height glass panels fixed between floors.

The ground floor covers a 6m-deep basement constructed using a bored pile wall.

Site space restrictions

A plan of the existing site is illustrated opposite showing the location of the new building relative to surrounding office buildings and roads.

Buildings A & B adjacent on the east side of the new building require access at all times for office workers.

After the erection of a close-boarded hording around the building perimeter, a space approximately 4 metres wide is left for material storage and building access.

Site logistic problems to be addressed by the contractor

- Protection of the public using Mount Street
- Providing access for staff using buildings A & B on the east side of site
- Siting of project offices and accommodation
- Siting of plant and equipment during the construction process
- Provisions for unloading materials and handling materials on site
- The effect of site restrictions on the construction methods
- Siting of subcontractors' requirements (offices/materials etc.)
- Provision of site security

Contractor's proposals

Protection of public

A covered walkway is to be provided forming part of the site hoarding along the west and south sides of the site.

Access to buildings A & B

The site hording is to be extended to provide an access way for office staff and visitors using these offices.

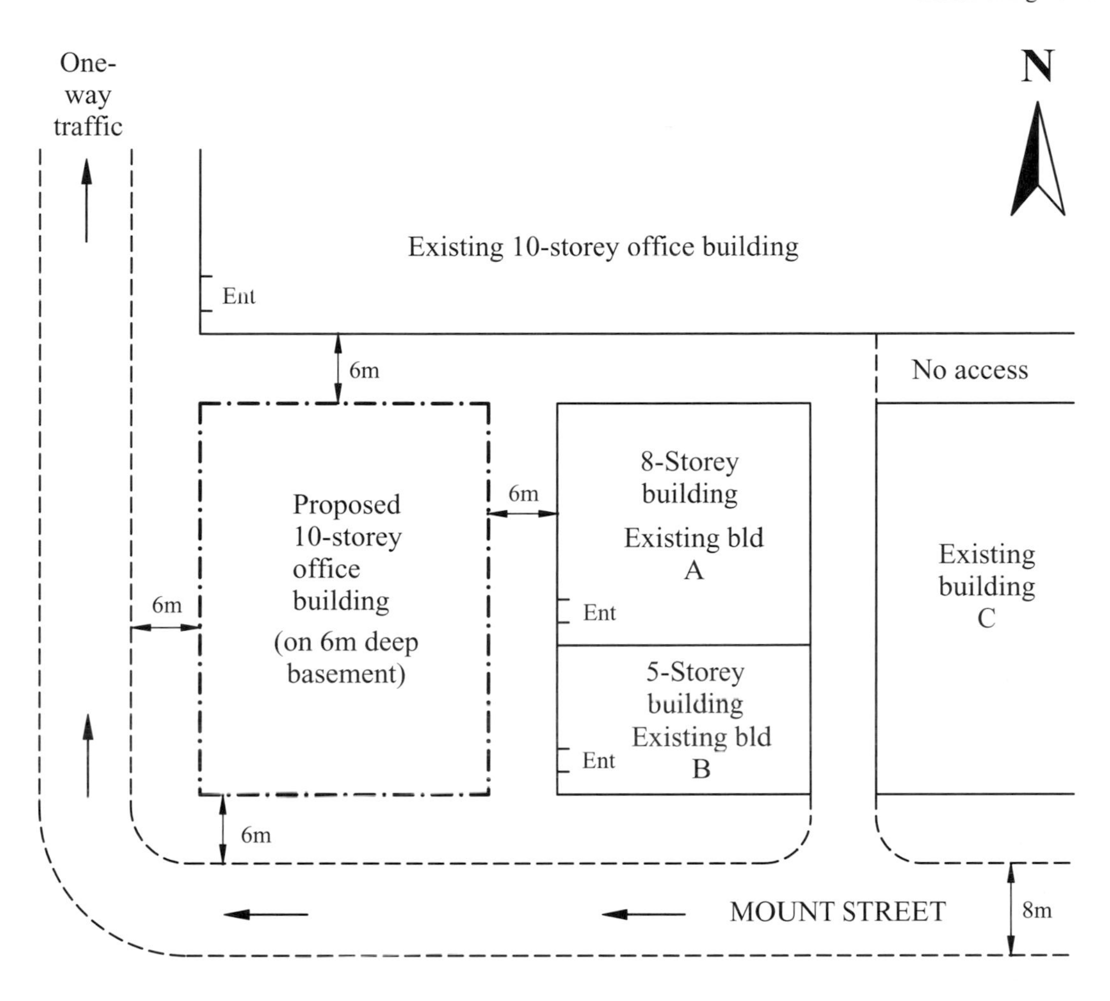

Site plan (existing)

Contractor's proposals (continued)

Siting of offices and accommodation

Due to space restrictions, the contractor's staff are to be located in rented office accommodation adjacent to the site.

On site, two mobile office units are to be situated in location 1 (see diagram opposite). These will be stacked as a tiered unit with operatives facilities' on the ground floor and the contractor's supervisory staff on the upper floor.

This will provide minimum accommodation requirements in order to supervise the basement construction and part erection of the steelwork.

On occupation of the basement roof (i.e. the ground floor slab) accommodation will be set up on part of the ground floor area.

A site security cabin will be provided in location 2 close to the entrance gates.

Siting of plant and equipment

A lufffing jib tower crane is to be located in the basement area at location 4.

An application has been made to the local authority to locate an unloading area on Mount Street at location 3. This has been granted, which enables delivery vehicles to be unloaded by the tower crane. It is also to be used for locating a mobile concrete pump when the building floor slabs are to be concreted.

The 4-metre space between the perimeter hoardings and the building will be used for locating mast climbers for erecting the curtain walling.

Provision of material storage areas / handling material

As the building frame is completed, the floor areas are to be used for material storage. Components will be scheduled for delivery on a just-in-time programme. Externally 'canti-boxes' will be used for loading materials into, prior to their being moved to the internal floor area.

During the basement construction the space arround the top of the basement will be used for fabrication of reinforcement and formwork.

The tower crane is to be retained for the majority of the project due to the handling of external glazing panels. The external curtain walling support is to be fixed from mast climbers working on each building elevation. The external glazing storey-height panels will be delivered on a just-in-time basis.

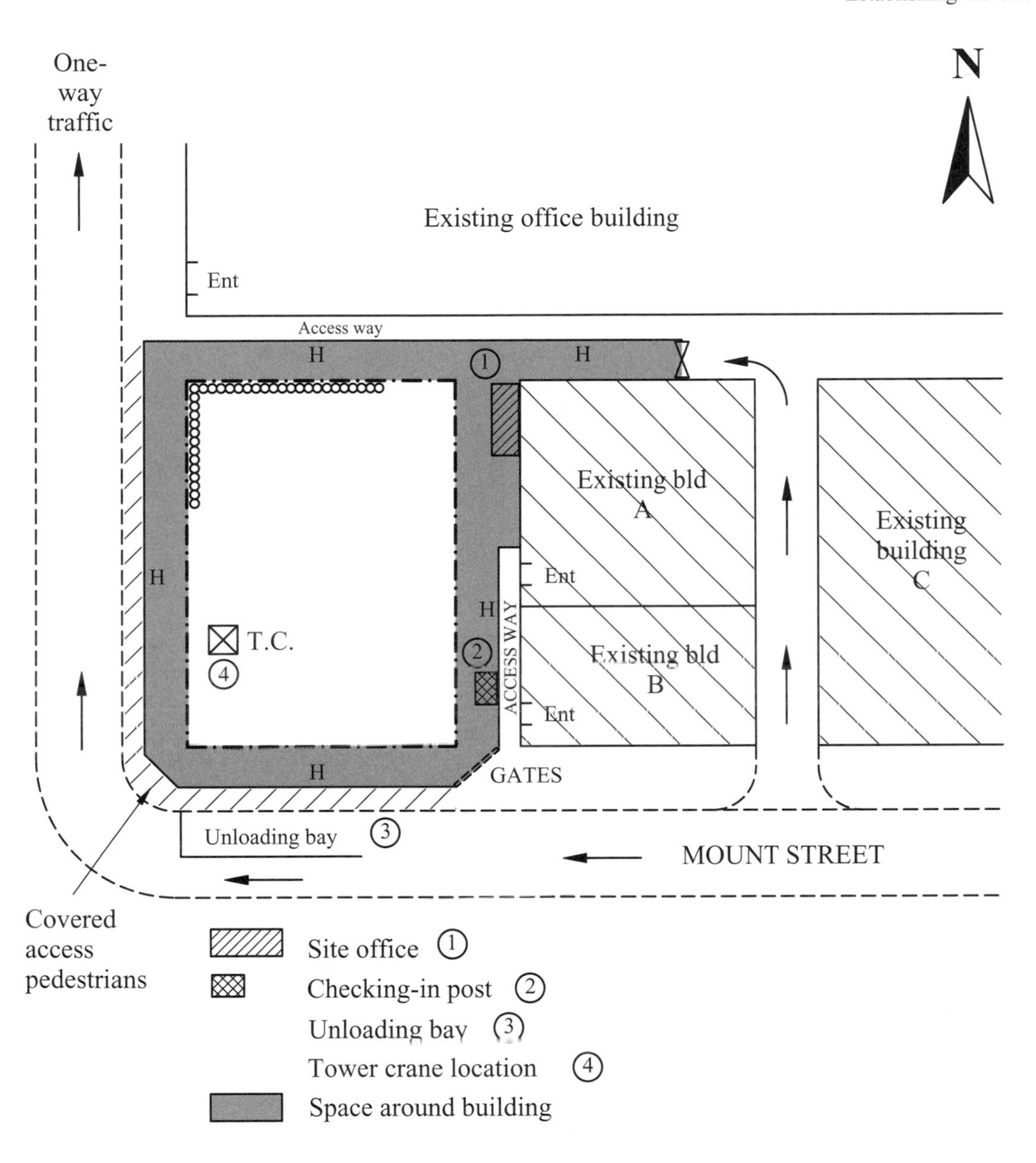

Site layout proposals

Site logistics office development

The image here shows the front elevation of the building under construction.

At the second floor level the building projects forward. This creates construction problems with the external enclosure work due to the restricted site space.

Location of tower crane

The luffing jib tower crane is to be located in the basement adjacent to Mount Street as shown on the proposed layout plan.

The four metres of storage space around the building can be used for the manufacture of form work and pre-fabricated capping beam reinforcement.

Concreting of the basement floor and ground-floor slab will be undertaken by mobile concrete pump, located in the site loading bay in front of the project.

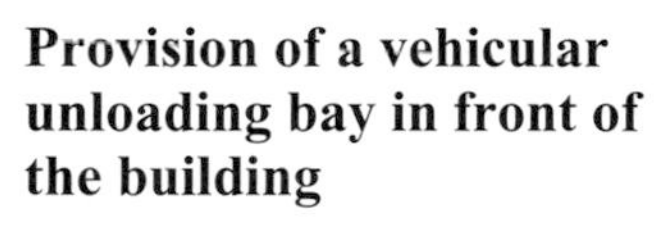

Provision of a vehicular unloading bay in front of the building

Due to site space limitations permission has been granted by the local authority to provide a vehicular loading bay in front of the building. The pedestrian access may be observed alongside it. The loading bay is also to be used for the storage of skips and pre-mixed mortar containers to be used during the works.

Access to the rear of the building

Access to the rear of the building, is at the rear of building A. This allows small vehicles to deliver site materials and light components. Pedestrian access has also been provided for the public around the rear of the site.

General view of the site from Mount Street, showing the proximity of the surrounding buildings.

The tower crane is erected in the basement area as work to the excavation proceeds.

Restricted space during the basement excavation involved 'haymaking' the excavated material from one machine to another. Refer to Excavation work Case Study 4.

General view of work in excavating the basement inside the bored pile wall as work proceeds. Capping beam constructed.

Site space restrictions can be readily observed. Site cabin (two tier) located alongside basement (location 1 on the site layout plan).

Concrete the upper floors, floor pour in progress. Concrete pump located in loading bay in front of building.

Note: Preparation of extended platform at fifth floor level from which to assemble mast climbers from floors 6 to 10. All the curtain wall supports are to be fitted from mast climbers. The full height external glass panels are to be handled into position by the tower crane.

The ultimate site facility: the wheel wash

Contractors are responsible for any damage or mud they deposit on public roads adjacent to the site. The local authority may contra charge them with the costs of cleaning the roads with mechanical road sweepers.

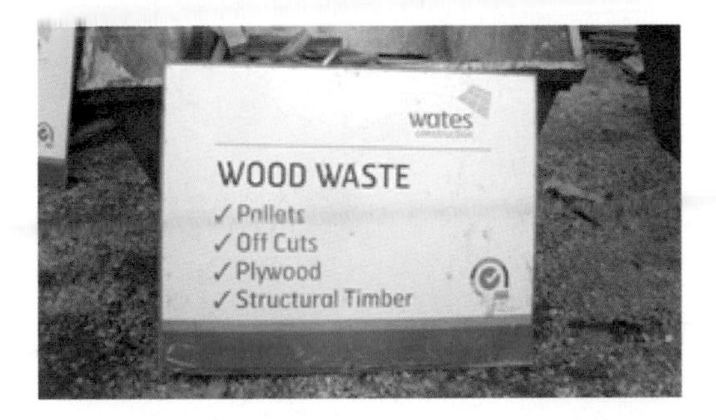

CHAPTER TWO

MATERIALS MANAGEMENT

Construction Practice. Brian Cooke

2.0 Overview

The site management responsibility for work on a construction project is first outlined. This is followed by a definition of types of waste generated.

The project manager's involvement in waste management is outlined, together with the duty-of-care requirements for checking and maintaining waste records.

The need to assess or identify waste streams is an essential requirement of any waste management system.

A series of images has been presented of good materials management practices and likewise some bad practices on site. It is far too easy to pass the blame for waste to another party. In construction all operatives and subcontractors should be made aware of their responsibility for controlling and reducing waste on site. The situations illustrated should help to give the student an insight into controlling waste on site.

Materials are an expensive commodity on site. The site manager must be aware of good and bad site practices. A waste separation policy should be encouraged and senior management must ensure sufficient monies are allocated in the contract preliminaries to enable this to be carried out. Strict implementation of the company's waste policy is essential.

A range of images relating to the use of skips is given, and the cost of providing skip facilities is also outlined.

Separation of subcontractor work and storage areas aids materials management

2.1 Responsibilities for materials management

Who is responsible for materials management on a construction project?

Is it the site management's responsibility?

- the site project manager
- the site construction manager
- the site supervisor
- the site subcontractor
- the site operatives

or is it:

- the company's senior management
- the company directors?

In fact, the responsibility for establishing policy lies with senior management.

Materials management practices must be encouraged from the top. The company is required to have a waste management policy and produce site waste management plans (SWMPs) for each contract. Consultants may be used to develop company policy and procedures and establish ground rules.

The company may develop their own procedures to suit their particular type of work. The careful management of materials will ultimately reflect in company profit margins.

The company has a duty of care and legal compliance regarding the storage and transfer of waste material. This is also to drive and encourage a reduction in waste generation on site.

Definition of waste types / catergories

Inert waste	Waste material that does not undergo physical or chemical change: soil, stone, clean brick etc.
Non-hazardous waste	Waste material that does undergo physical or chemical change but is not hazardeous: metal, wood, plastics etc.
Waste	Waste is anything which is discarded. Surplus material that is to be used or reused on site is not classed as waste material. Surplus material should be stored separately from waste.
Hazardous waste	Materials such as asbestos Chemicals contained in the ground Contaminated ground etc. Dangerous chemicals and materials Disguarded masic tubes Tile adhesives

Responsibilities of the project manager for waste management and material control

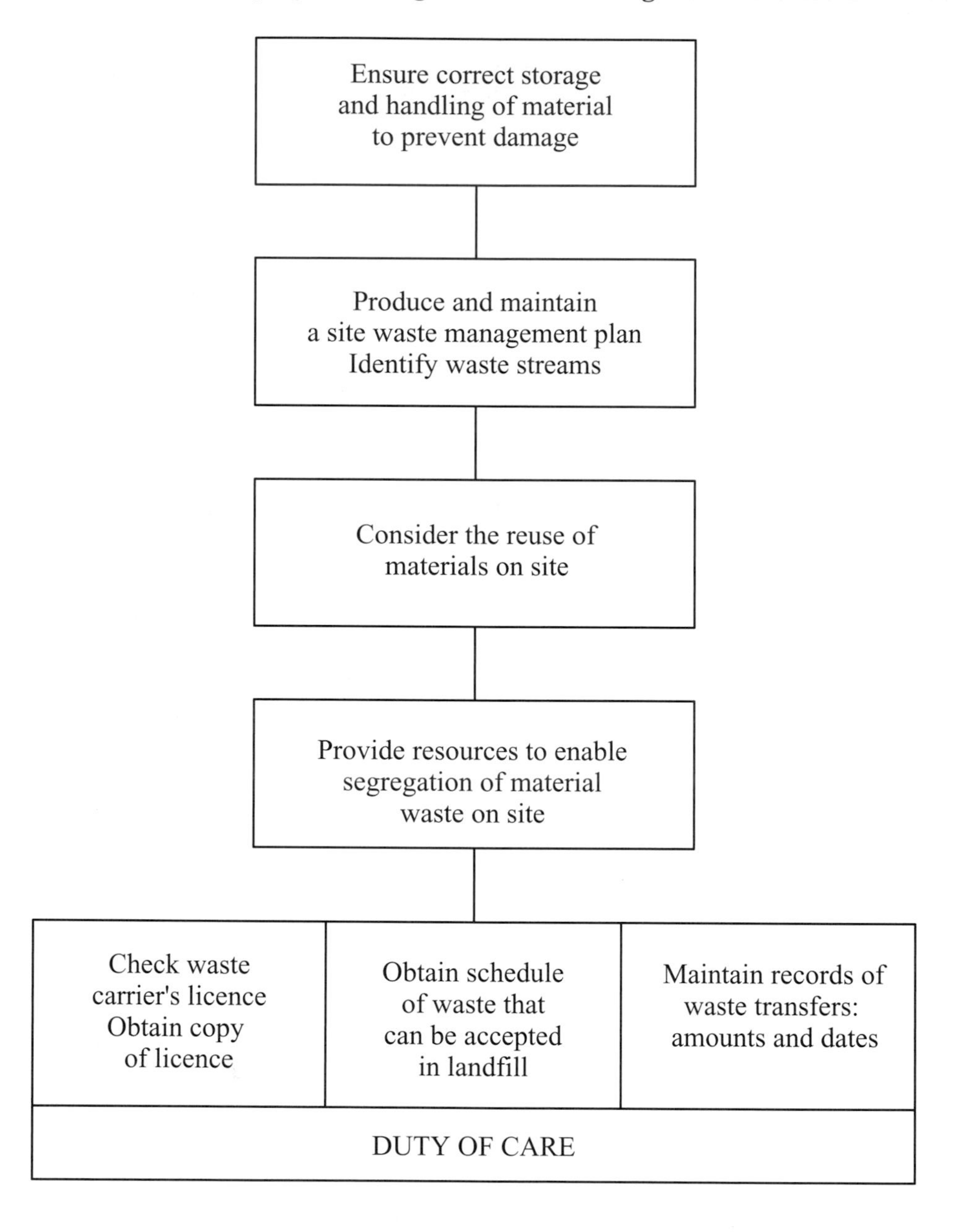

2.2 Creating waste streams

Identify waste created by each stage of work, i.e. identify waste streams generated by undertaking the site operation or stage of work. Make operatives aware of site waste policy during site induction.

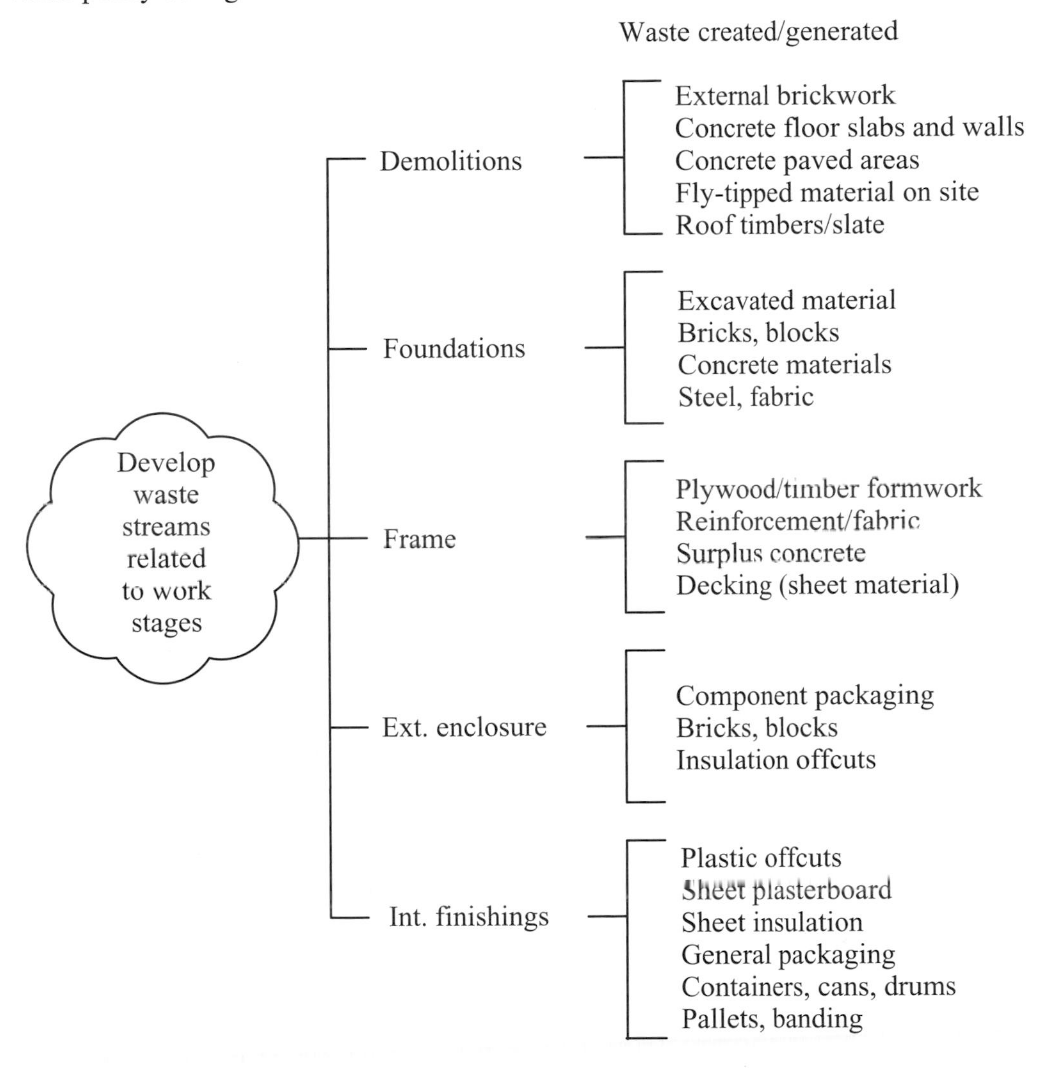

Assess quantities of waste generated by each waste stream or stage of work.
Propose how waste is to be dealt with and/or disposed of.
This assessment forms part of the site waste management plan.

2.3 Good site practice observations

- Maintain a tidy site. Implement a policy of checking major work areas at the end of each day or phase of work–completion of bathroom and of fit-outs.
- Provide prepared hardstanding areas for brick, timber, roof sheeting etc. and ensure that materials are either palleted or stored clear of the ground.
- Cover and protect material likely to be damaged by rain/exposure.
- Use storage containers (containing racks etc.) and storage bins for the collection of materials around the site.
- Maintain a paleting policy where timber pallets are strategically placed around the work area. A 'pick up a brick' policy may be encouraged.
- Encourage the separation of waste into separate skips or marked disposable waste sacks. Subcontractors are to be responsible for moving their own waste to the skip points.
- Ensure that operatives and subcontractors are fully aware of company materials policy. Include at the induction stage.
- Do not allow subcontractors to 'take over'. Where all trades are sublet' this situation may easily occur (refer to bad practice situations). Consider contra charging subcontractors with the wastage they produce or charge them daywork for cleaning up.
- Use of tool-box talks and the daily notes whiteboard or hazard board to report on site practices. Adopt a name-and-shame approach to continual offenders.
- Provide dedicated work areas for specific operations, allowing space for material storage within the area (e.g. steelwork erection)

The responsibility for control of materials ultimately lies with the project manager. Checks should be applied to key materials such as bricks, blocks, ready-mixed concrete etc. Whereby the bill quantities are matched with ordered quantities and used quantities i.e. a material reconciliation process.

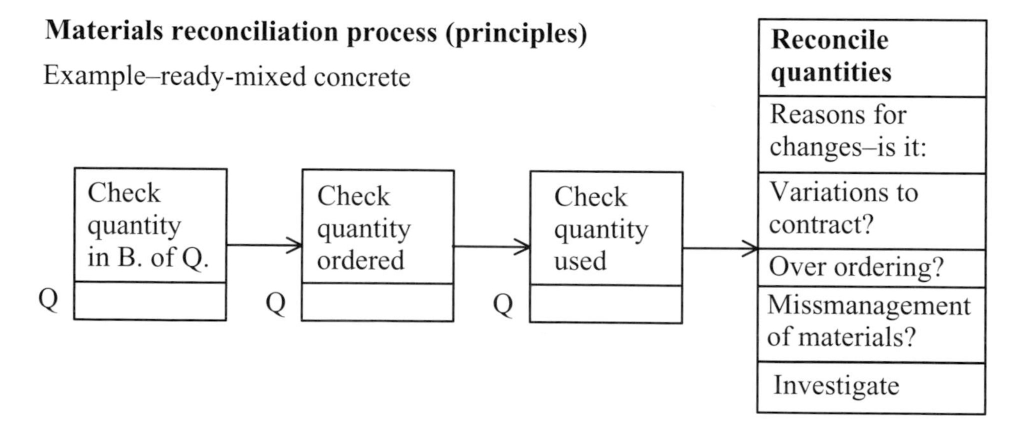

A tidy site is a clean site

Use of storage bins for fittings

Storage racks for scaffold materials

Use of storage containers

Good site practice

The key to successful materials handling is to provide a site layout plan. This should show the planned main storage areas. The company must allow sufficient monies in the contract tender to provide for these services.

A site layout plan should be prepared which clearly indicates where materials are to be located relative to the building. Areas for brick storage should be pre-prepared prior to the delivery of the goods–a 'palleted, policy may be implemented on the site. Provision should be made to protect materials from the weather.

Blocks stored on a prepared hardstanding

Good site practice

Profiled sheet stacking clear of the ground

Organised storage area

Skip content signage

See-through storage bins

Timber pallets for placing blocks on in localised areas

A tidy site is a well-organised site

2.4 Bad site practice observations

One can find poor material management practices on most construction projects. Blatant practices by individual operatives cannot be controlled or ruled out.

Incidents observed include:

- stacking insulation blocks too close to a roadway–when the road becomes muddy the blocks get splashed with mud and become unusable
- unloading block or brick pallets by pushing them over or hitting them with a shovel to unband them
- poor handling and stacking of such items as timber roof trusses, sheet material and timber and not storing materials clear of the ground
- complete lack of thought in the organisation of storage areas and total disregard to the cost of materials and components
- cutting careless waste of sheet insulation, plasterboard and plywood materials
- failure to store cement in dry conditions; failure to clear up areas around site mixing plants, leaving pre-mixed mortars to set; over-ordering of concrete materials

Solution:

Make subcontractors include the material costs in their quotations.
Make them accountable for the waste they create.
Photograph the havoc they create and contra charge them for it.
Name and shame on the site hazard board.

A disorganised site
No consideration prior to delivery of material
There is only the site manager to blame

Mass disorganisation–a real shambles

Palleted blocks stacked too close to muddy access road

The easiest way to unload blocks –push them over!

Roof trusses stacked against the scaffold

Site waste management practice

Is your site really this bad?

The following sequence of photographs illustrates the management's attitude towards waste on a speculative housing project. The developer/contractor has obviously lost control of the management of waste on site. One may blame these occurances on the site management team or on the attitude of subcontractors towards materials management. It is too easy to blame it on the subcontractors. The responsibility can only lie with the site manager, and his superiors (the contracts manager) who obviously allows the situation to continue.

Perhaps it is simply due to the company having no waste management policies, or no senior managers responsible for implementing it.

Bad site practices

Disorganised storage area

Remains of a cement pack

Cutting-waste problems with sheet insulation

It can't get any worse than this!

Surely not more photographs of materials waste?

These observations are from contracts undertaken by small medium-sized contractors in 2008/9.

The company had a site management waste plan but little emphasis appears to be given towards the implication of it. On these sites, no waste separation of materials takes place. Everything is thrown into the skip, and left to the skip company to dispose of it and sort out.

The requirements for contractors to prepare site waste management plans (SWMP) was introduced in April 2008. Contractors are required to forecast the amount of waste generated by site operations (or site waste streams as the project moves from one stage to another). In the above situations this is clearly not being addressed.

It seems that many small contractors have no waste management plans in place at all.

2.5 Use of skips on site–waste separation practices

Good waste management practice

Photographs showing the use of skips on site
for the collection and disposal of materials

Use of waste packaging skip and designated area for sealant/adhesive containers.

Separation of timber materials in crane handleable skip.

Separate materials into separate skips: Timber, plywood, metal ducting, bricks etc.

Waste separation
Good practice

It is important that operatives and subcontractors are awarc of the company policy on waste control at site level. Waste management practice on site should form part of the site induction process.

Signs highlighting waste management practices will aid implementation of the site management waste plan.

The separation of waste materials on site must be encouraged by the adequate signing of skips. Subcontractors may be allocated specific waste disposal areas for specific materials and timber items.

The costs of providing and disposing of site skips (plus the value of materials carted away in them) is difficut to assess.
The actual cost can amount to some £500 per skip.

Use of skips on site

Non-waste separation
Other practices

Contractors often have a waste management policy but fail to implement it on site.

It is too easy to place all rubbish in a single skip and leave to the skip supplier to sort or separate the material out once the skip is returned to the depot.

Many of the large companies contract their waste management to a single natural waste company. In these cases the skip practices illustrated would never materialise on site. It is up to the contractor's site waste management team to ensure that site waste is properly controlled.

2.6 The waste disposal process

Premier Waste (UK) Ltd

General view of waste transfer station

COMPANY PROFILE

Premier Waste has been established some fifteen years and has established links with major contractors in the North West region.

The company endeavours to provide a direct service to major contractors in the north-west region, acting as both waste carrier and waste managers. Major companies serviced include Carillion, Kier North West, Taylor Woodrow and Morgan-Ashurst.

Advice is provided to project teams on bespoke waste stream analysis reporting, SWMP and the segregation of waste material on site. Each project/site is issued with a report on the percentage of waste recycled each month (refer to analysis sheet indicating the recycled volumes). The aim is to recycle 90–95% of mixed waste in order to ensure that a minimum amount is sent to landfill sites. A layout plan of the waste transfer station operated by Premier Waste is illustrated.

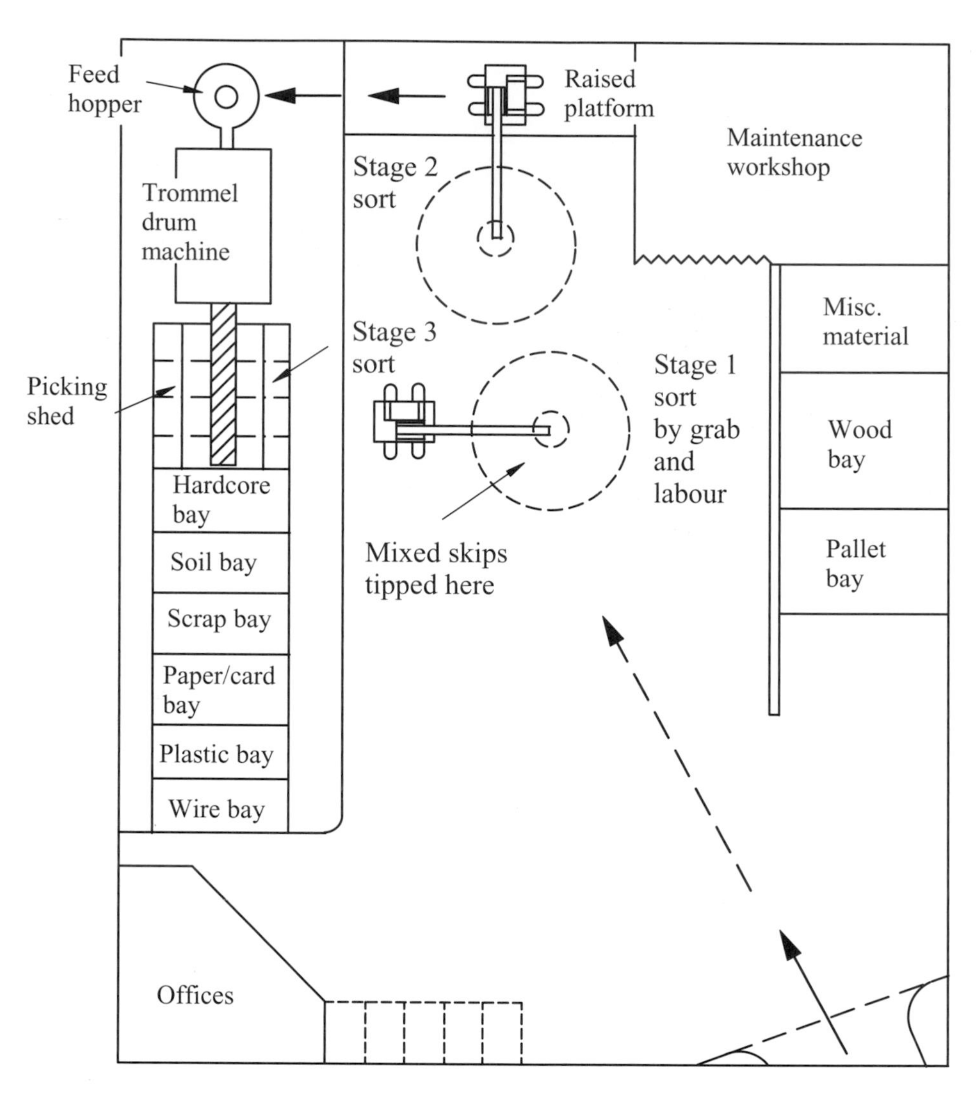

Layout of waste transfer station

Stage 1 sort –By labour and excavator fitted with rotary grab

Stage 2 sort –Material moved within reach of grab/excavator on raised platform feeding hopper of trommel drum

Stage 3 sort –Eight labourers in picking station feeding sorted materials into bins below station

The waste disposal process

Images around the works

Stage 1–Sort with rotating grab

Premier Waste skip vehicle

Stage 2–Sort feeding hopper to trommel drum

Rotary grab machine on platform

Trommel drum machine–outputing fines

Stage 3 Sort in picking station (eight labourers)

The sorting process

Skip costs–Mixed waste skip sorted at transfer station

Costs are indicated for a 20m3 capacity skip containing different types of waste.

Waste type	Hire/transport	Disposal
General waste	£110	£60 per ton
Wood waste	£225	Included
Inert waste (soil/concrete)	£195	Included
Scrap metal	£110	Credit for content
Paper and card	£110	Credit for content
Plasterboard	£110	£60 per ton
Hazardous drum	£75/drum	Included

The cost of landfill tipping of active waste is subject to a government landfill tax of £48 per ton–plus the tipping fee per load charged by the specific tip. A yearly increase of £8 per ton has been set by the goverment for the next 5 year period. Non-active waste such as soil is used to cover material on active waste tipping areas. The tipping of non-active waste is not subject to the government landfill tax.

Analysis of recycled material provided to clients

The report indicates the analysis of data produced by Premier Waste on the recycling of waste on a site-by-site basis. A pie chart showing recycled volumes is also available.

Site specific report

Recycled Volumes of Waste

Site/name ____________ Site Ref. _________ Month ________ Year ______

Waste analysis of 24 skips at the above project.

Euro ID	Waste Catagory	Recycled c.m.	Land fill c.m.
Euro code references/ waste catagory	Bricks	22.9	1.0
	Wood	22.7	3.4
	Plasterboard	16.6	0.2
	Concrete	11.4	0.6
	Plastic	6.3	1.2
	Paper/cardboard	5.4	1.4
	Soil and stone	4.6	0.3
	Metals	4.1	0.2
	Packaging	3.0	0.6
	Other	2.3	0.1
	Asphalt and tar	1.5	0.1
	Tiles and ceramics	1.5	0.1
	Municipal waste	0.6	10.3
	Insulation	0.2	3.8

Recycled 82% Recycled volume 102.2 Landfill volume 23.3

The company aim to achieve a recycle percentage of 90–95.

CHAPTER THREE

DEMOLITION
AND
EXCAVATION

3.0 Overview

Demolition work

Various methods of undertaking demolition work are summarised in general terms.

Four demolition case studies are included illustrating a range of construction methods.

- French seven-storey building
- Garage and forecourt
- Four-storey block of flats, including method statement extracts
- Two-storey steel-framed workshop block

The importance of method statements is outlined, including the presentation of written and tabular method statements.

Excavation work

Excavation plant

A range of excavation plant is illustrated, together with appropriate applications. Hazards relative to excavation work are outlined. A wide variety of applications are shown for tasks involving mini-excavators and general excavation plant.

Five case studies are included relative to excavation works:

- Deep pit involving a battered excavation
- Large basement
- Pile cap bases together with an illustration of the sequence of work
- Six-metre-deep basement on a restricted access site
- Basement to a bored pile wall

3.1 Demolition methods

Demolition techniques depend on the nature of the building being demolished. The risks to the public and operatives involved in the demolition process, and the location of adjacent structures need consideration.

Method statements and risk assessments must be submitted, to and approved by, the main contractor prior to demolition works being carried out. The local authority building control must be kept informed of proposed commencement dates and hazards likely to affect the public.

Methods of demolition include:

- Piecemeal demolition by hand
- Demolition by wire rope pulling
- Demolition by machine, excavator etc.
- Demolition by demolition ball (balling machine)
- Demolition by hydraulic pusher arm
- Demolition by explosives
- Use of impact hammers and nibblers
- Deliberate controlled collapse by weakening the structure

The building to be demolished may be of traditional construction, or a building with in-situ or precast concrete floors. A steel-framed building poses different problems compared to an in-situ or precast concrete frame.

Every demolition project raises its own construction problems. An initial building survey will be necessary in order to provide a realistic method statement.

Demolitions

All demolitions and dismantling works are covered by CDM. Part of the planning work should include the provision of method statements.

Piecemeal Demolition

Piecemeal demolition is demolition done by hand, using hand held tools and is sometimes preliminary to other methods. When the structure has been demolished by hand to a height of 10 metres, then conventional heavy equipment may be used.

Controlled collapse and pre-weakening of the structure

Deliberate controlled collapse involves pre-weakening the structure of the building. This involves removing key structural members so that the remaining structure collapses under its own weight.

One method of removing key structural members is overturning by wire rope pulling. In this method, wires are attached to the main supports which are pulled away using a heavy tracked machine or vehicle.

3.2 Demolition case studies
Case study 1-Seven–storey flats (France)

Demolition of seven-
storey block of flats

The seven-storey block is of precast concrete crosswall construction incorporating precast floor and wall units

General excavator
fitted with 'nibbler'

A large hydraulic tracked back actor (high-reach excavator) fitted with a 6 metre demolition arm and attachments for breaking up floor and wall panels

'Nibbler' in action

The excavator is fitted with a 'nibbler'. Nibblers use a rotating action to snap brittle materials such as concrete and masonry. They can crush up floor panels once they are located at ground level.

Removal of material
from site

After demolishing and breaking up the floor and wall panels, the material was loaded into 10 c.m. waggons using a medium-sized back actor/excavator.

Case study 2–garage demolition

Demolishing steelwork to petrol station

General view of garage and forecourt work

Excavating to remove underground concrete

Demolition of garage and forecourt

The project involves the demolition of a garage building, steel canopy and concrete forecourt area. A decision was made to recycle as much material as possible–hence the decision to use an on-site crushing plant. The crushed material is to be recycled for filling under site roads and paved areas for a new housing development.

Removal of large piece of concrete

Site crushing plant

Case study 3–four-storey flats

Picture of typical block

Stripping of structural roof trusses
Strip out flats of reusable materials

Grab for lifting and loading timber and concrete material

External arm for the pusher attachment to excavator

Description of project

Demolition of six blocks of four-storey deck access flats. Brick external walls precast / in-situ floors and decks.

After stripping the roof by hand in order to recover timber components, the building was demolished using an excavator fitted with a pusher arm. An excavator fitted with an open grab attachment was used to separate materials.

Case study 4–Two-storey workshop

Here comes 'the beast'

Tracked hydraulic machine fitted with 2.5 c.m. bucket and 360° pivoting grab arm

Pivoting grab arm at work

Demolition of modern two-storey steel-framed building, brickwork to first floor level, with precast concrete first floor. Internally the building has partitioned classroom areas at both ground and first floor.

Stage one demolition includes the soft strip-out–i.e. removal of skirtings, doors, partitions and handrail materials to staircase areas.

All concrete to the ground floor slab and brickwork was crushed up for reuse as hardcore on site.

The large machine simply tore the building apart with the rotary grab arm.

Excavator demolishing, crushing and bending steel frame

Duration:	Soft strip-out	7 days
	Demolition work	9 days
	Crushed material for reuse	4 days
		20 days

All crushed brick and concrete reused as hardcore on site

3.3 Method statement extract for demolition works

METHOD STATEMENT EXTRACT FOR DEMOLITION OF FOUR-STOREY FLATS

Method statement headings–written format

General –Scope of works
–Programme
–Site establishment
–Traffic management
–Perimeter protection

Demolition works
–Soft strip of internal fixtures and fittings
–Demolition of the structure

Soft strip of internal fixtures and fittings

Prior to soft stripping (removal of timber skirtings, suspended ceilings, floor coverings and partitions) the operatives are to remove agreed external windows at each floor level (so that they can throw the stuff out) to establish designated drop/loading zones. The structure is to be soft stripped of all fixtures and fittings using traditional methods and hand tools.

Skirting boards and door frames are to be removed using pinch bars and hammers. Suspended ceilings are to be removed by operatives working from mobile access platforms.

Carpet coverings are to be removed by shovel. Large carpets are to be cut into strips for ease of handling.

Stud partitions and plasterboard coverings are to be removed separately and moved to the designated drop zones.

The works are to progress on a floor-by-floor basis. The material in the drop zone will be separated into convenient sizes (may be done by a 360° excavater fitted with a large rotary grab attachment). The material can then be separately loaded into skip containers for disposal.

Demolition of structure (i.e. brick external walls and floors)

The walls are to be pushed inwards into the internal floor voids by a 360° excavator fitted with a push arm.

The material will then be loaded into 10/12 c.m. lorries for removal from site. Additional breaking up of the wall and floor material may be undertaken by the excavator. Separation of all crushable material will be an ongoing process with stockpiles created at suitable points (either awaiting loading into lorries or into a crushing machine if available).

All materials are to be removed to licensed disposal points via sheeted lorries with full documentation being supplied upon completion of the works.

Activities with a risk/hazard potential

- Heavy plant movement
- Moving lorries and equipment
- Working at height
- Manual handling
- Hand breaking
- Mud on the road
- Glass around work area
- Fall objects
- Dust
- Noise
- Slips, trips and falls
- Uncontrolled collapse
- Access and egress
- Services

Personnel protective equipment to be used

- Safety helmet
- Safety boots
- High visability clothing
- Coveralls
- Safety harness
- Ear protection
- Eye protection
- Dust masks
- Respirators

3.4 Method statements–format

Construction method statements are an essential part of contract procedures. Their purpose is to convey proposed methods of construction to the client. Subcontractors are also required to prepare them before their operations can be commenced on site. Method statements form part of the safety file on site.

The method statement may be presented in a wide range of formats as there is no standard method of presentation–each one to his own.

Formats include:

- Written method statements relating to single site operations (task or site specific).
- Global method statements covering all site operations (ie. method statement and exit plan)–this is a pre-contract method statement which covers the main construction elements and handover/sequence forecast.
- Tabular method statement, which is operational and often relating to programmed site operations– this format requires a separate risk assessment.

The headings used on the above method statments are illustrated.

Written method statement

- Description of operation stating how the work is to be carried out
- Identification of labour and plant resources
- Risk involved in undertaking the operation.
- Control measures to reduce the risks
- Monitoring compliance and personnel responsible for implementing statement.

Descriptive list of the main stages of work or building elements

–Establishing site

–Foundations work

–Frame and roof

–Building enclosure

–Building services provisions

–Building finishes

–External works

–Exit plan - phasing and hand over sequence

Tabular method statement

Typical format (headings on landscaped sheet, left to right)

- Operation: number and description
- Quantity (approximate quantity of work, if applicable)
- Proposed method of undertaking the work
- Labour resource
- Plant resource
- Notes on operation period, work sequence etc.

A separate operational risk assessment must be provided with this format due to space restriction on the sheet.

Risk assessments are often provided separately prior to an operation commencing; once again each company tends to develop its own format.

An example from a tabular method statement for undertaking site clearance and foundation work is illustrated on page 64.

Examples of method statements are illustrated in various chapters.

A long reach hydraulic excavator with excavator bucket fitted
A pusher arm for demolition work may also be attached

Foundation works			Method Statement	Broker Foundations Date. October 28th 2010		
Op. No.	Operation	Quantity	Proposed method	Resources		Notes
				Labour	Plant	
1	Excavate to reduced level ave 1m deep	600 c.m.	Excavate to reduced level using large hydraulic back actor machine–load 10 c.m. capacity waggons and cart to tip off site.	1–banksman	4–lorries (hired)	Output 200 c.m. per day Duration 4 days
		60 c.m. crushed brick	Allow for forming hardcored access road for vehicular access for lorries			
2	Excavate foundations 600mm wide ave. 1m deep	120 c.m.	Excavate with medium-sized back actor, load material into large dumper and deposit in fill areas on site	1–banksman	1–Exc. 1–4 c.m. wheeled dumper	Output 40 c.m. per day Duration 3 days
3	Concrete to strip foundations	30 c.m.	Place concrete by direct discharge from ready mixed-concrete waggons alongside trenches. concrete to be placed as excavation work proceeds	3–labourers	1–vibrator	Allow for one pour per day to suit excavation progress

3.5 Excavation work–excavator types

Excavators are classified as small (mini excavators), medium or large; this is relative to their function, bucket size and output capacity.

Small excavators

Small excavators are classified as mini, midi or compact excavating equipment. Other types of micro excavators are designed to pass through 750mm wide door openings. The excavator may be rubber or metal-tracked. The majority of machines have a pusher blade and a back hoe excavation arm at the rear. It is not possible to refer to manufacturer's model references as one would become totally confused with machine types. A wide range of attachments are available which include mini-breakers, compactors, earth drills, hedge cutters and handling equipment.

Medium and small excavators (Hitachi)

Manufacturer's include JCB, Caterpillar and others. A wide range of bucket sizes and attachments are available for both the tracked and rubber-tyred versions. This includes breakers, tine, forks and compactors.

Machine outputs vary from 10 to 40 c.m. per hour depending on machine size and the type of excavation work being undertaken. Access for loading vehicles will also influence output.

Method statements and risk assessment

Before excavation work commences, proposals should be submitted to the contractor/ client in respect to the proposed methods of working.

The method statement may be submitted in a written or tabular format depending on the contractor's or subcontractor's preference.

Headings used in a written statement of proposed construction methods may include:

- operation (description of task)
- proposed method
- labour and plant allocated/resources
- assessment of hazards
- control measures proposed
- responsibility for approval

The construction method should reflect the time allocated for the task relative to the construction programme.

The method statement will require approval by relevant parties before work commences.

Hazard assessment

Likely hazards:

- movement of plant - Excavator and lorries being loaded
- persons falling into excavations
- plant falling into excavations
- excavation collapse/earth collapse
- access to the excavation or excavated area

Control measures:

- provide experienced banksman
- provide barriers around top of excavation both for operatives and plant
- provide adequate earth support techniques (sheet piling, trench sheets, trench boxes, drag boxes–consider angle of repose when considering a battered or sloping excavation)
- provide safe egress and exit routes from basement– i.e. ladder or scaffold access via stairs and ramps
- highlight daily hazards on the site hazard boards
- provide adequate directory and warning signs

Subcontractor's responsibilities

Most excavation work is undertaken by subcontractors. They should be made aware of their responsibilities with respect to the provision of method statements and hazard assessments.

Reference–A number of textbooks relate to construction methods and the application of practical method statements.

- Cooke & Williams (2008) Third Edition, Construction Planning Programing and Control, Wiley–Blackwell.
- Illingworth J.R. (2008 reprint) Second Edition, Construction Methods and Planning, Spon Press.

Excavation applications

Levelling site

Use of large hydraulic back actor for grading and levelling surface of excavated area ready to receive stone fill under large floor slab area.

Site clearance

Excavate to reduced levels with large hydraulic back actor load 10 lorries for removal of excavated material from site.

Stone spreading

Use of medium-sized excavator for spreading stone and fine surface dust for a small roadworks project.

Demolition work

Use of medium-sized back actor for demolishing three-storey brick building. Loading demolition material into 10/12 c.m. lorries and cart to tip.

Spreading stone fill under floors

Small JCB mini excavator fitted with pusher blade and larger bucket.

Excavate service trenches for foundations

Small Kobuto mini excavator fitted with trenching bucket for work excavating site service ducts.

Inner city town work

Clearing out raised flower bed areas with hydraulic back actor. Front bucket used for carting away shrubs from flower beds.

Hydraulic breaker fitted

Used for breaking up tarmac area to roadway or break out road kerbs for removal to tip.

Versitility of the mini-excavator

Stripping the site of turf and topsoil

Forming spoil heaps adjacent to the work

Excavating foundation trenches average 600mm wide by 1m deep

Handling concrete kerbs and paving flags. Breaking up paved areas

3.6 Excavation works–case studies
Case study one–battered excavation works

Six-metre-deep excavation in bolder clay strata. Excavation undertaken in two stages–note change in batter line on photograph. Access ramp formed to excavate second stage and provide access to lower level. Approx. 1000 c.m. excavated in a five-day period.

Case study two

Excavation of large basement

Description of project

The work involves constructing a 4m deep basement within a bored pile wall. Total volume of bulk excavation to be removed approx 8000 c.m. plus grading bottom of excavation to form floor level.

On completion of the excavation work a 300mm layer of limestone was laid to form a piling mat. Two hundred CFA piles were then formed to provide the foundations for the basement floor and main building works.

The following photographs illustrate the methods of undertaking the work.

A temporary ramp was formed into the excavation for vehicular access.

Main excavation work undertaken using two large back actors loading 10–15 c.m. lorries for removal to tip off site.

Output:

2 no. large excavators
(400 c.m. per day × 2) 800 c.m./day
Duration of task: 10 days

Trim base of excavation:

1 no. medium-sized back actor loading dumpers and deposit on site
Duration of task: 3 days

Case study three
Excavation of pile cap bases

Description of project

Excavation work to a number of different-shaped pile caps average 1.2 m deep. Exposed piles ready for machine cropping. Two, three and four piles per pile cap are illustrated. The sequence of work involved in the formation of a piled base is illustrated.

The following illustrate the work involved in excavating the pile caps.

Excavator type used: medium-sized tracked back actor with labourer trimming the excavation as work proceeds. Due to the nature of the ground, no earth support of perimeter formwork was required. After placing the steel reinforcement and positioning bolt boxes the bases were concreted.

EXCAVATION WORK
FORMATION OF A PILE CAP

Stages of construction

(1) Insert piles

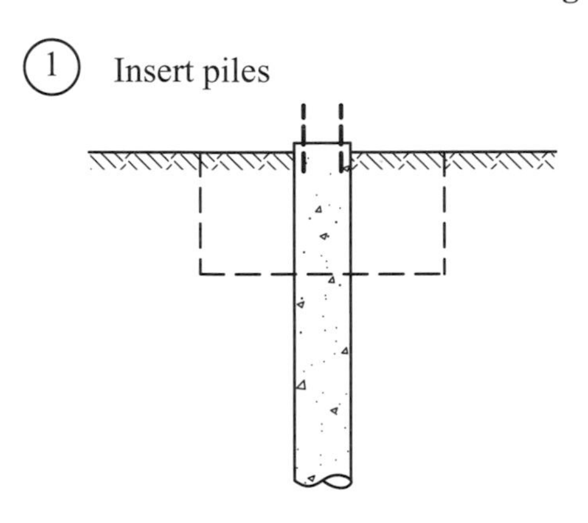

(2) Excavate pile cap

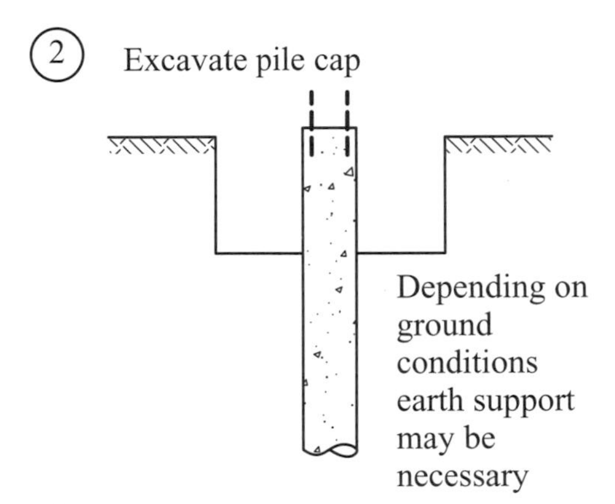

(3) Blind pile cap

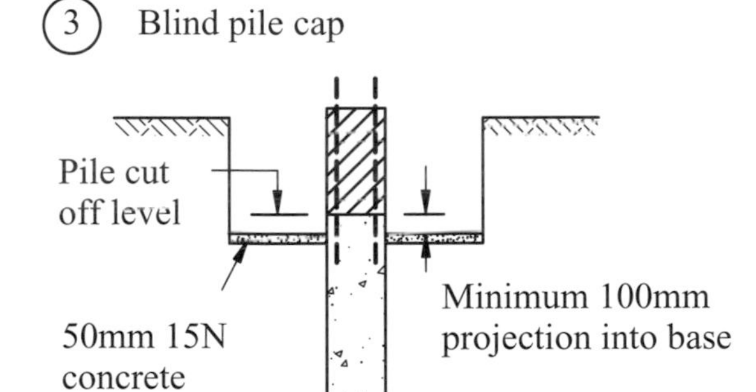

(4) Cut off piles

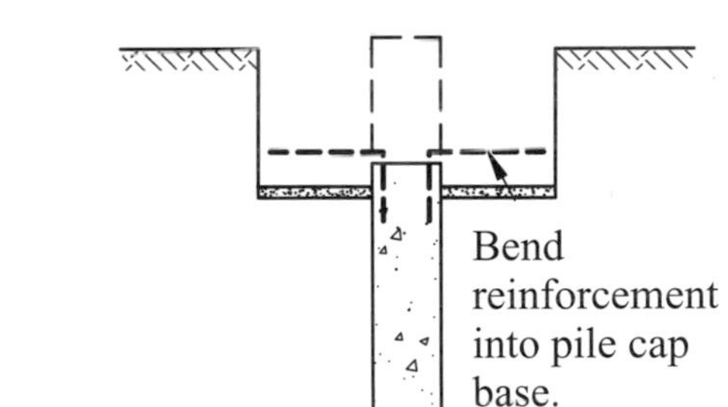

(5) Position/suspend holding down bolts in position

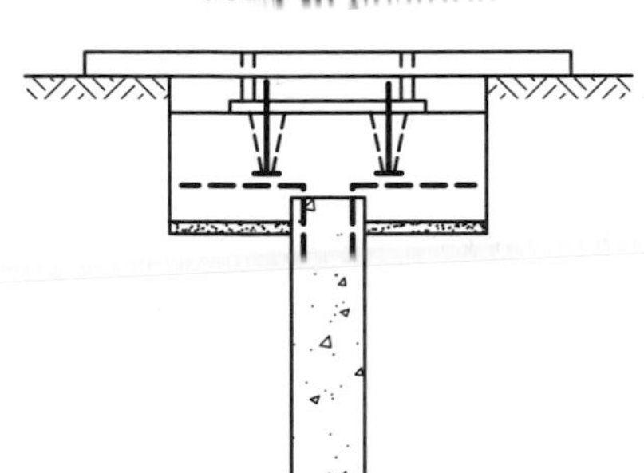

(6) Concrete base
Clean out bolt holes

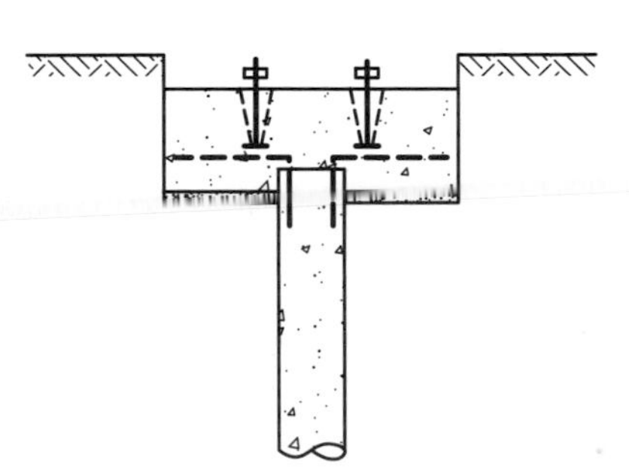

Preparation of a pile cap

Excavate large base and expose 900mm diameter piles

Mark exposed pile with cut off level

Hydraulic pile cropper for piles up to 1m diameter

Completed pile with base concrete blinded

Excavation work–case study four

Excavate to six metres deep-bored pile basement

Description of project

The site has restricted access and limited on-site space.

The 25m square basement is 6m deep.

Excavation is being undertaken using two medium-sized back actor machines.

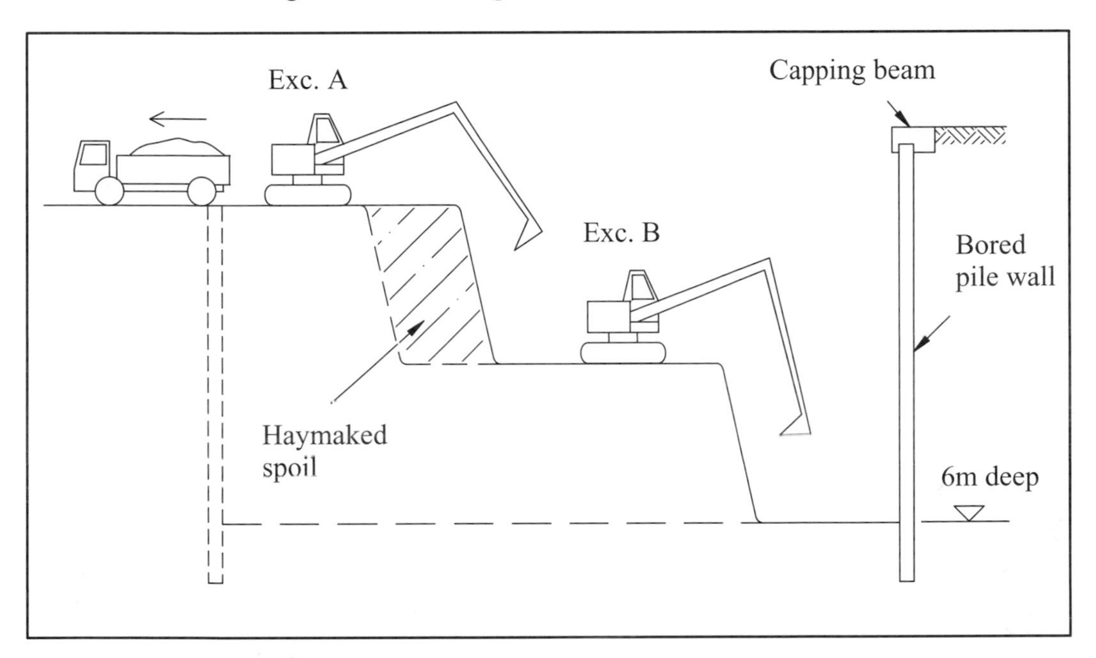

Excavator B 'haymaking' excavated soil to excavator A. Extremely restricted site space around top of bored pile wall.

Excavation work–case study five

Basement excavation

Excavate basement average 5m deep

Extensive activity shown with one excavator working at floor level, pushing up material to a second excavator located at ground level.

Small yellow mini-excavator levelling stone to form piling mat at basement floor level. Continuous flight auger piles to be driven to support basement slab and foundation bases to two seven storey-apartment blocks.

Excavate basement–expose piles

Capping beam complete

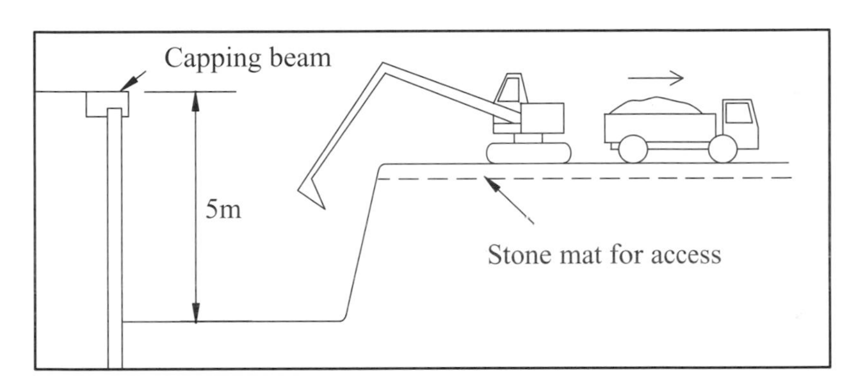

Excavate to basement

General excavation–bored pile wall exposed

Clearing excavation to final levels

CHAPTER FOUR

FOUNDATIONS AND PILING

Construction Practice. Brian Cooke

4.0 Overview

This chapter covers all types of foundations suitable for both domestic and commercial buildings. Sketches of various foundation types are carefully linked to photo images from construction projects.

Strip, deep strip and pad foundations are first outlined together with related ground floor construction where appropriate. The application of trench blocks is illustrated.

Pad foundation types illustrate the connection of the frame to the foundation i.e. protruding holding-down bolts, column kickers / or provision of tapered pockets to receive precast columns.

Five types of raft foundations are illustrated together with a case study for constructing a simple downstand beam raft for domestic housing and a 24m diameter raft slab for a ten-storey precast building.

Displacement piles are illustrated relating to precast piling using both a large rig and mini pile arrangement. A mini-piling system (Bullivant) is further illustrated together with a precast prefabricated system of ground beams.

Replacement piles are illustrated on a multi-storey building project using wet hoppers in conjunction with ready-mixed concrete. A further example is included when forming 400mm diameter CFA piles.

An example is included relating to the use of vibro-replacement techniques (stone column method) for a single-storey pub complex. Sketches of the foundation layout plan for the vibro piles are shown.

Many of the popular construction textbooks cover piling techniques in detail.

These include:

Barry's Advanced Construction of Buildings (Stephen Emmitt / Christopher Gorse) - First edition 2006

Building Construction Handbook (Chudley)

Building in the 21st Century (Robert Cooke) - First edition 2007

Construction methods and planning (JR Illingworth) - Second edition 2000

It is hoped that the use of images from a digital camera will encourage students to present better-illustrated projects relating to aspects of foundation construction.

4.1 Foundation types–strip foundations

Types:

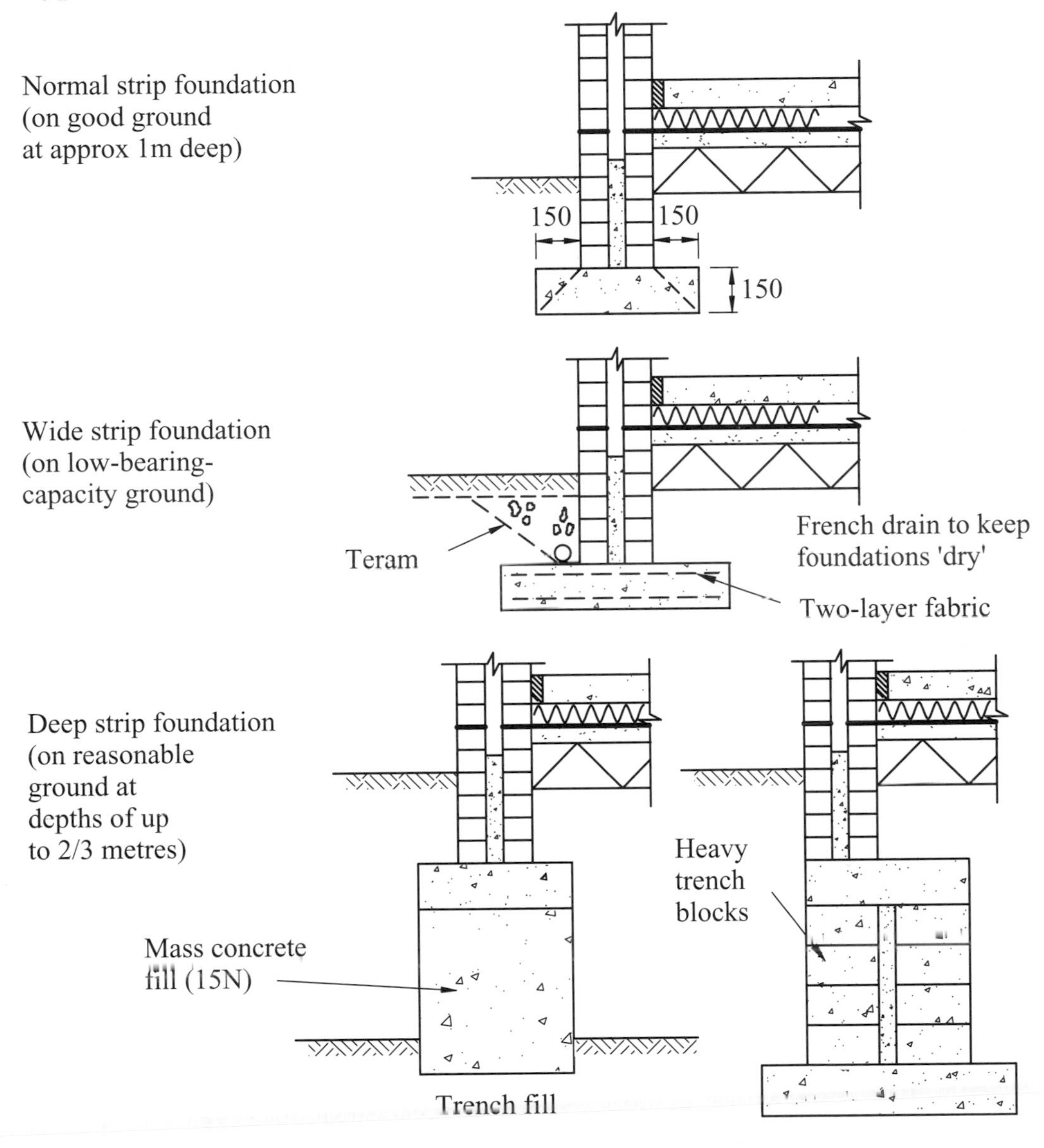

Various forms of strip foundations are illustrated depending upon ground conditions (establishing the bearing capacity of the soil at foundation level)

Strip foundation

Strip foundations shown average 900mm deep for a single-storey building.

Three courses of external blockwork are shown with drainage entry points (note concrete lintels over openings). The load bearing in-situ concrete ground floor slab is laid on 300mm of compacted stone fill.

Strip foundation using trench blocks

An example using Durox foundation airated blocks. Block size 350 x 310 x 215 mm.

Three courses of Durox foundation blocks. This method speeds up the construction process and eliminates the need for separate leaves and the use of wall ties and cavity fill.

Tarmac topblock Durox airated blocks used for the foundations to support a beam and block suspended ground floor slab.

Strip foundation

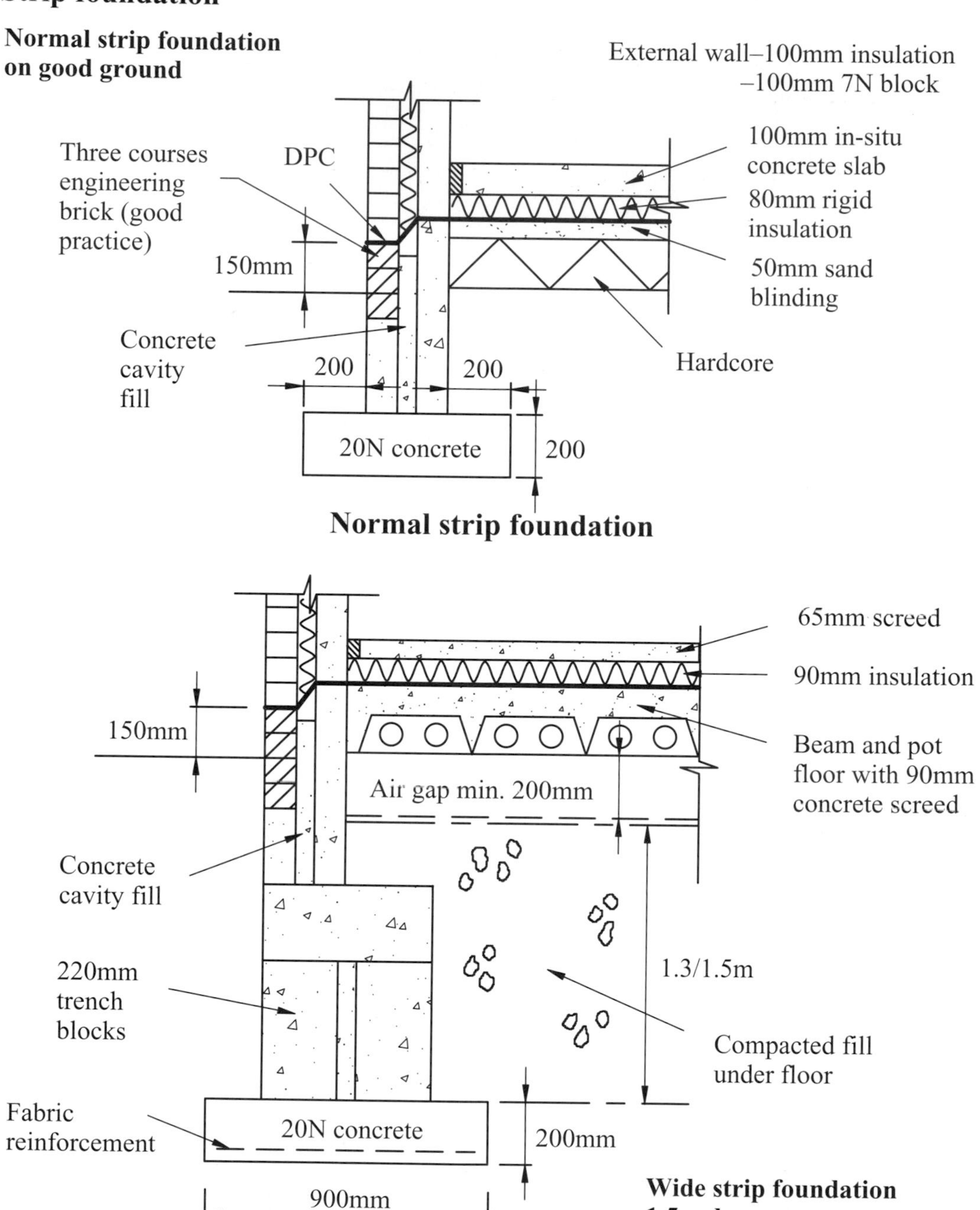

Normal strip foundation

Deep strip foundation

Deep strip foundation

Foundations illustrated for a low-rise residential housing project. Oversite excavation to a depth of 1.5m prior to forming 900mm wide-strip foundation. Walls to ground floor level constructed of heavy trench blocks incorporating three courses of engineering brick below DPC level.

Compacted earth fill under floors up to 1.2 metres deep. Beam and block suspended floor constructed to form ground floor slab.

Deep strip foundation

Earth fill being compacted using small plate vibrator. Drainage to ground floor slab being laid in plastic drainage ware. Note three courses of engineering brick below DPC level.

Precast I beams positioned to receive infill blocks to form suspended ground floor slab.

Prefabricated strip foundation on piles

Layout of precast ground beams supported on bored piles capped off with precast or in-situ pile cap (by Roger Bullivant Piling and Foundation Engineers).

The precast ground beams support the exterior wall and suspended beam and block floor. Speedy and efficient approach to foundations on brownfield sites.

Junctions between precast ground beam connected with in-situ concrete joint.

4.2 Foundation types–pad foundations

Purpose: to spread a point load from a column into the ground.

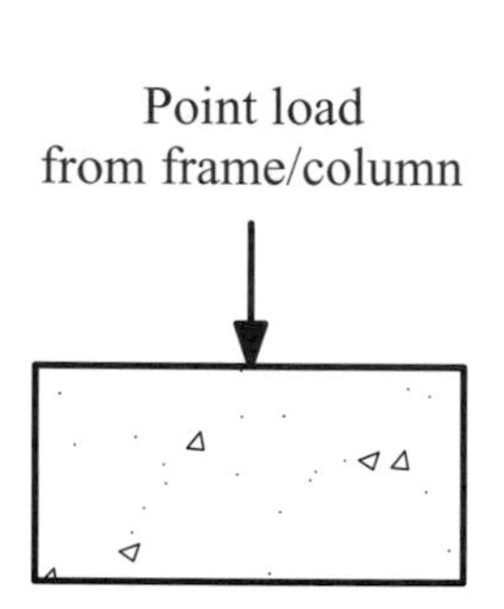

Mass concrete pad

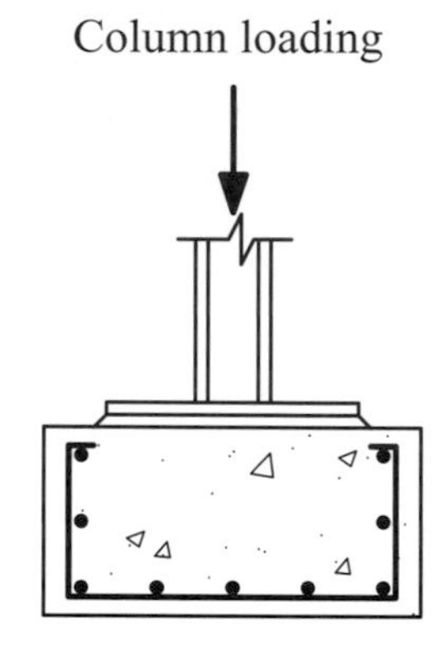

Reinforced pad foundation

Connecting the building to the foundation

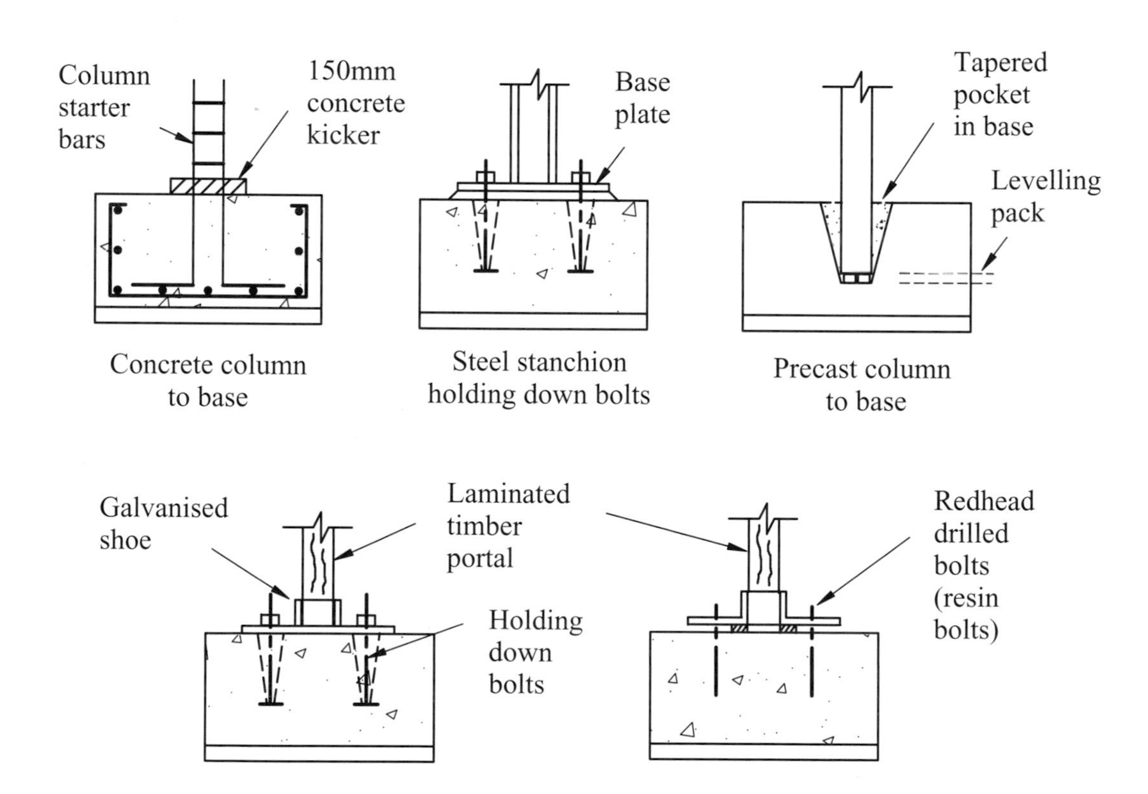

Isolated pad foundation base showing completed reinforcement and column starter bars.

Lowering prefabricated reinforcement cage into position prior to fixing formwork.

Completed square pad foundation with adjacent base ready for concreting.

Complete pad foundation bases with projecting column reinforcement 150mm high concrete column kicker to be formed on base.

Fabricated base plate for holding down bolts to a steel stanchion. Tapered sleeves to be fitted around bolts prior to concreting the foundation base.

Pad foundation

An image can express a complete construction sequence

Image illustrates:

- Excavation completed to pile cap
  Piles cropped and base blinded
- Pad foundation reinforcement prefabricated for square and rectangular foundations
- Bolt boxes to be prepared, reinforcement positioned and pad foundation to be concreted

Refer to Case study 3 (Chapter 3)

Pad foundations and ground beams to a framed building

Pad foundations may be bearing directly on the ground or supported on piles depending on ground conditions.

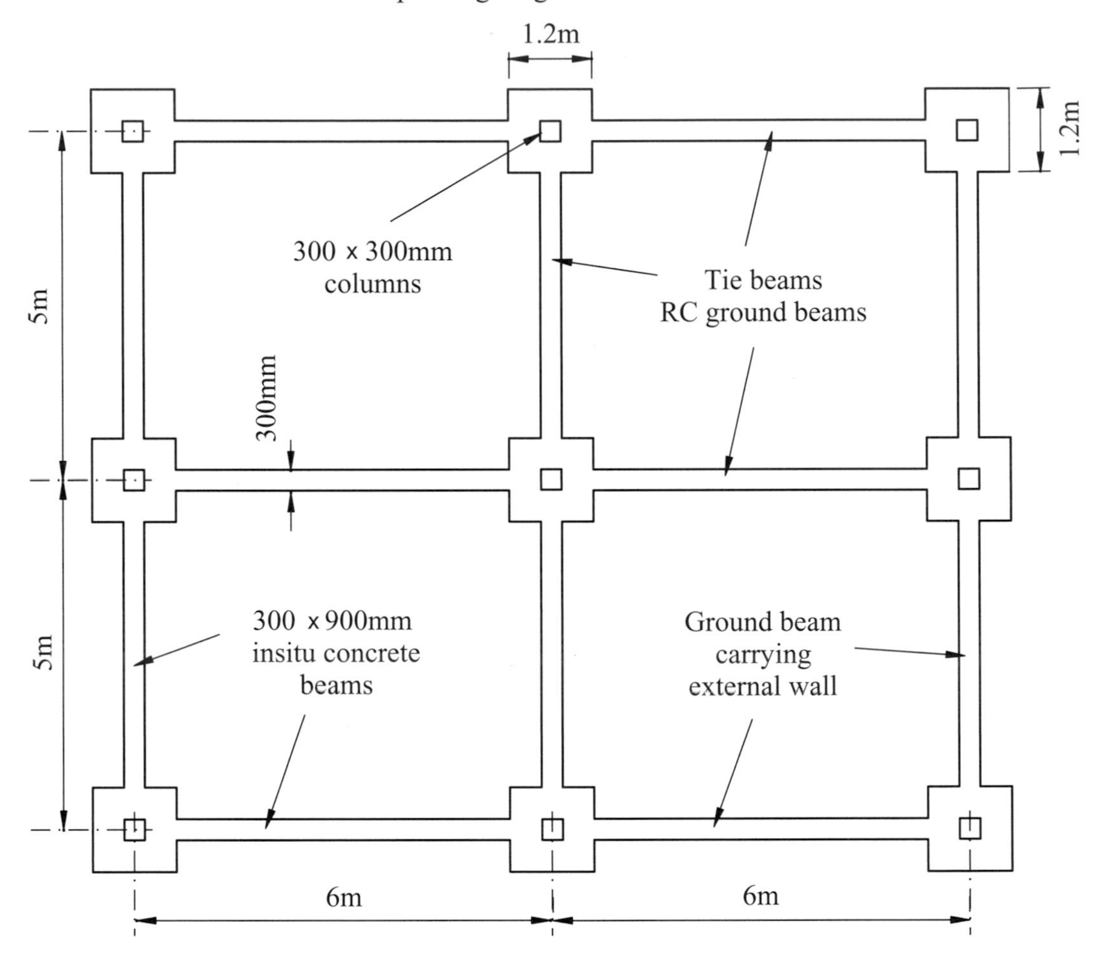

Foundation plan

This type of foundation layout suits a variety of forms of construction, i.e. column and beam frame, flat slab or a structural steel frame.

4.3 Foundation types–raft foundations

Flat slab raft (simple flat slab) 250mm thick concrete slab on 400mm of granular fill. Paving around building to form protection to foundation from frost.

Flat slab raft with toe beam (Wimpey-type raft)– practical, economic form of construction.

Downstand beam raft– perimeter beam (not an economic solution) due to extensive amount of formwork involved.

Beam and slab raft with downstand beams– load from crosswalls (as per circular building case study). May be supported on piers.

Cellular raft or hollow core raft to reduce weight of foundations.

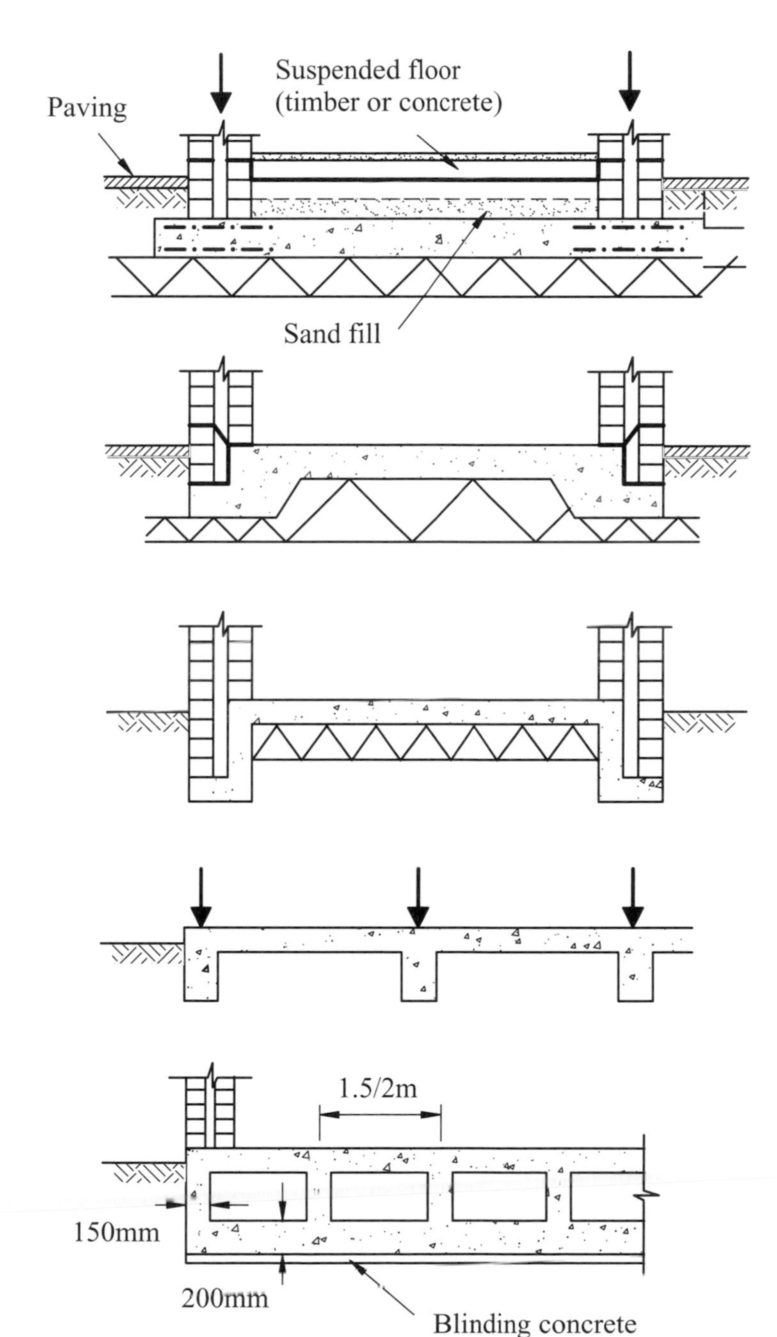

Raft foundation–flat slab raft with toe beam

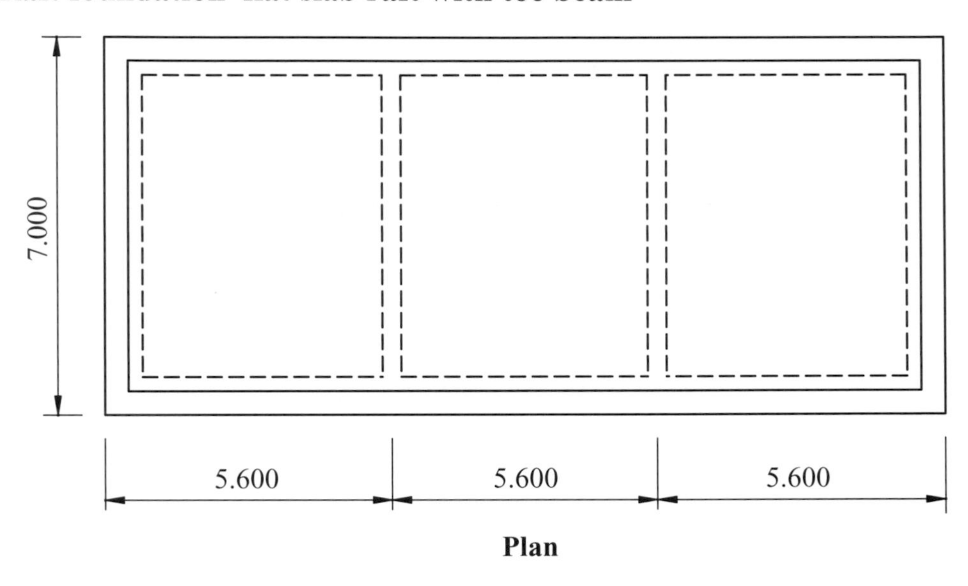

Plan

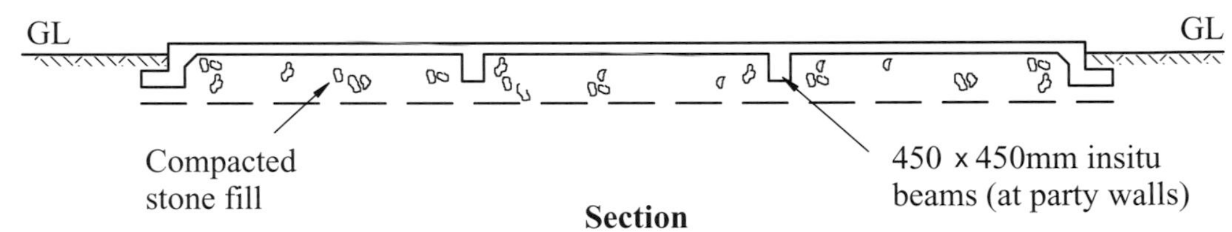

Section

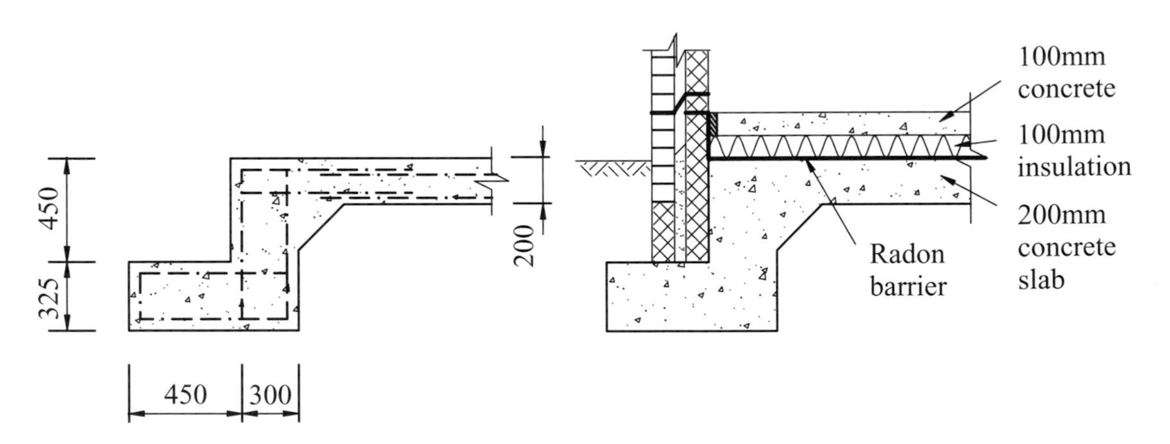

Raft Toe Beam **Foundation Section**

Sequence of work forming downstand beam

Stage one

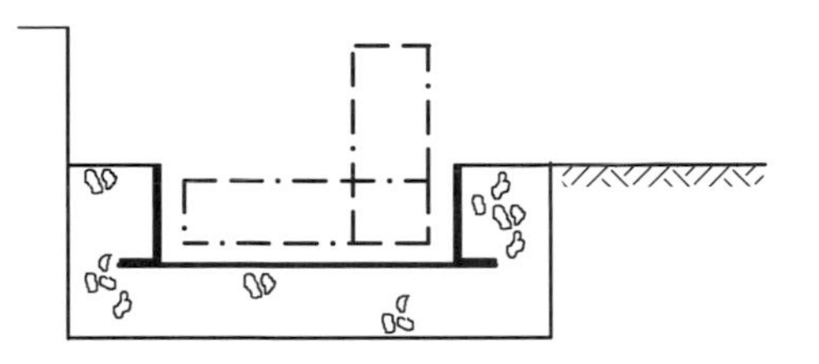

Excavate, lay stone fill, position reinforcement and Pecofill edge formwork (left in)

Stage two

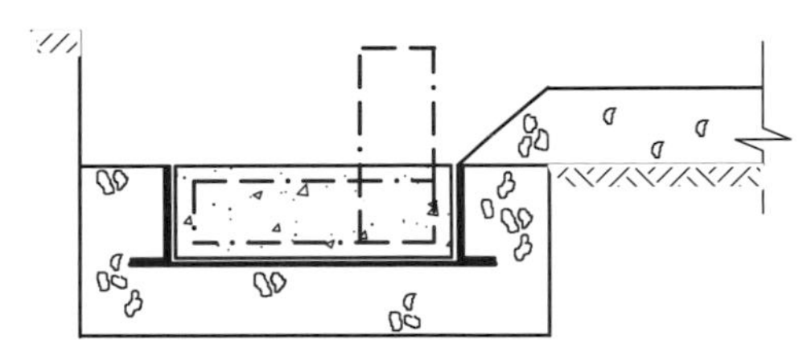

Concrete the foundation toe, form up stone fill under main raft slab

Stage three

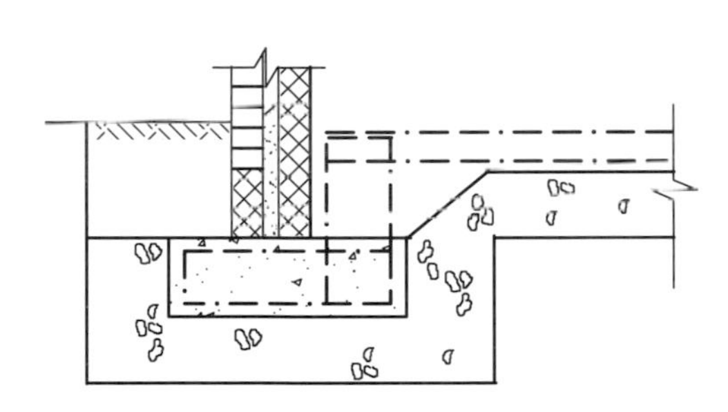

Construct external cavity wall on toe to provide retainer to edge of raft slab. Fix reinforcement to floor slab

Stage four

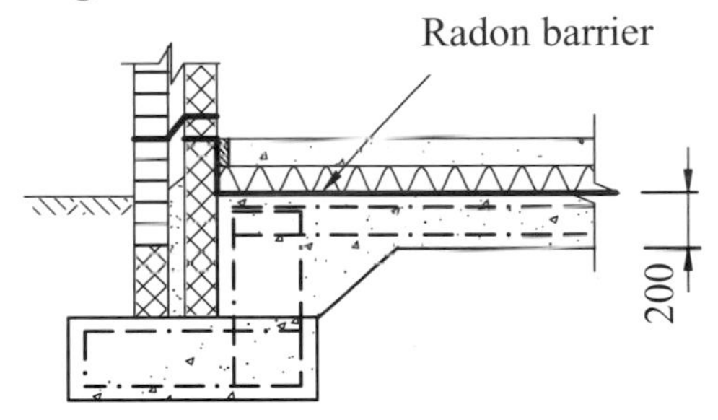

Place in situ concrete thickening to raft and ground floor slab

This designed raft slab is suitable for use on domestic house foundations for a single dwelling or terrace of linked houses.

Stage one and two in forming the raft

Oversite excavation to raft complete. Laying and compacting limestone under perimeter toe beam and party walls

Fixing toe and upstand beam reinforcement

Pecofil permanent foundation formwork in position to raft toe. External stone backfill to be completed prior to concreting

Stage three and four in forming the raft

Concreting of the raft toe beam after backfilling to outer edges of formwork. Concrete placed with excavator bucket

Cavity wall built on raft toe
(in lieu of formwork to edge of raft)

Concrete raft slab

Raft foundation

Completed raft slab with edge formwork stripped

Description of project

The project includes the construction of two ten-storey residential blocks. The main structural frames are of precast crosswall construction and are to be erected at the rate of one floor per week per block.

The project value is approximately £20 million with an overall contract period of 50 weeks.

A case study based on this project is included in Chapter 20 of Construction, Planning, Programming and Control - Authors Brian Cooke and Peter Williams.

Raft foundation

Each raft slab is 24m diameter and 400mm in thickness. The slab is supported on fourteen downstand radiating ground beams from the central core area. The ground beams are further supported on 50 bored piles of 300mm diameter. The site was previously a contaminated site containing an extensive gas plant. A methane barrier is incorporated in the raft slab.

The contract was completed on programme and budget. Details of the crosswall construction are given in the building frames section where, an extract from the precast programme is included.

Ground beam layout under raft

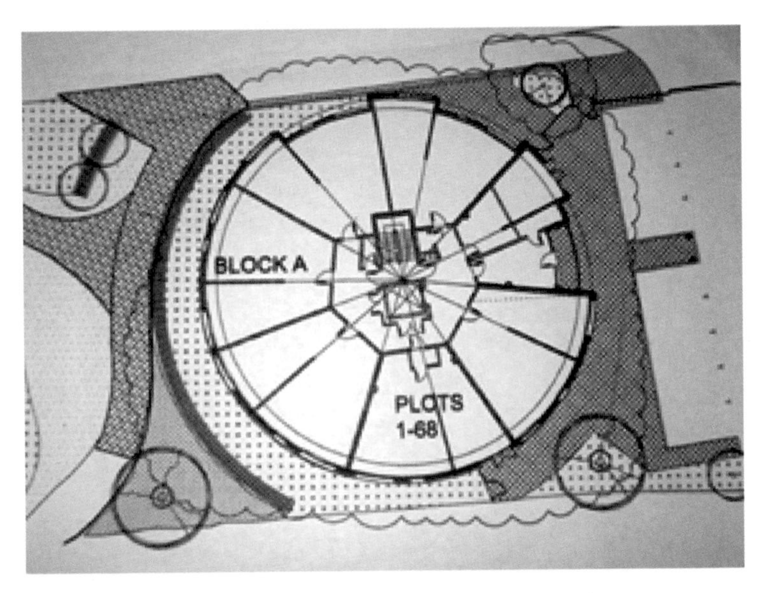

Plan of 24m diameter building. Each floor contains seven residential units based on crosswall construction

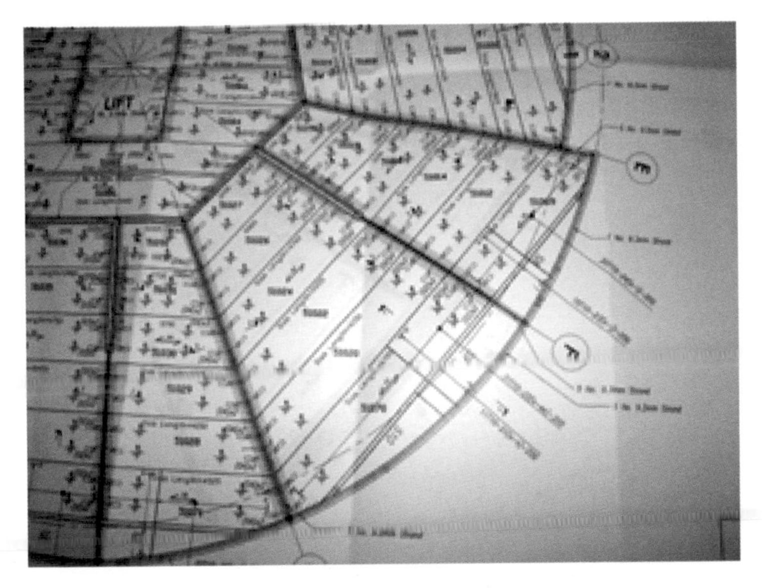

In-situ concrete downstand ground beams supported on bored piles

General view of ground beam under raft slab–radiating from central core area. The core area contains a central lift and internal staircase

Gas membrane laid under the raft slab

Formwork complete to raft perimeter slab. The raft slab was concrete pumped as a single 140cm pour

Concrete pumping in progress. The gang are due for a hard day's pour. Gang size: four labours plus two finishers.

4.4 Piling in construction

Some personal observations and experiences on piling.

There are many good construction technology textbooks which cover the principles and applications of piling techniques, so why attempt to re-invent the wheel?

- Emmitt S, Gorse C. (2006), Barry's Advanced Construction of Buildings. Blackwell, Oxford.
- Warren D.R. (1996), Civil Engineering Construction. Macmillan, London.
- Chudley R & Greeno R, (2006), Building Construction Handbook.
- Holmes R, (1983), Civil Engineering Construction.
- Illingworth J.R. (2nd Edition,1996), Construction Methods and Planning, E and F.N. Spon

In 1956 I was a site engineer on a large industrial project where the piling system being installed failed to meet specification requirements.

The Simplex-driven piling system was being undertaken on site. This was a displacement pile system using a steel lining tube and iron shoe.

On withdrawal of the lining tube (during concreting) the lining tube was raised above the concrete level, allowing the ingress of water to the pile shaft– this defect is referred to as 'necking' of the pileshaft.

Consequently the piling system failed and additional piles had to be driven at the piling contractor's expense,

Since these early days in civil engineering, experience has been gained of a wide variety of piling techniques across the spectrum: i.e. bored, driven, diaphragm and bored piled walls. Perhaps I am now entitled to describe some of the modern practical applications by a series of sequential photographs.

Piling illustrations

Four piling examples are illustrated:

- Displacement piles-precast driven, used on a development of five three-storey blocks
- Mini-piling system for lift shaft base (driven–displacement piles)
- Replacement piles using continuous flight auger for the foundations of four-storey steel frame building
- Vibro-displacement system of improving the bearing capacity of weak soils by the formation of stone columns using a soil displacement process.

Piling in construction

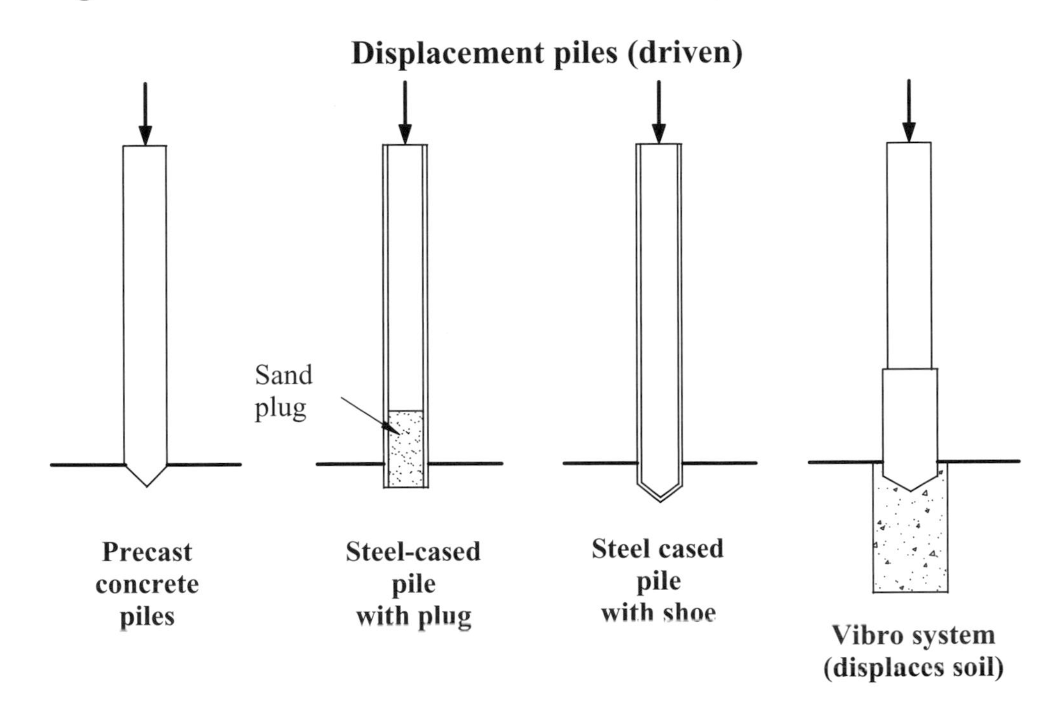

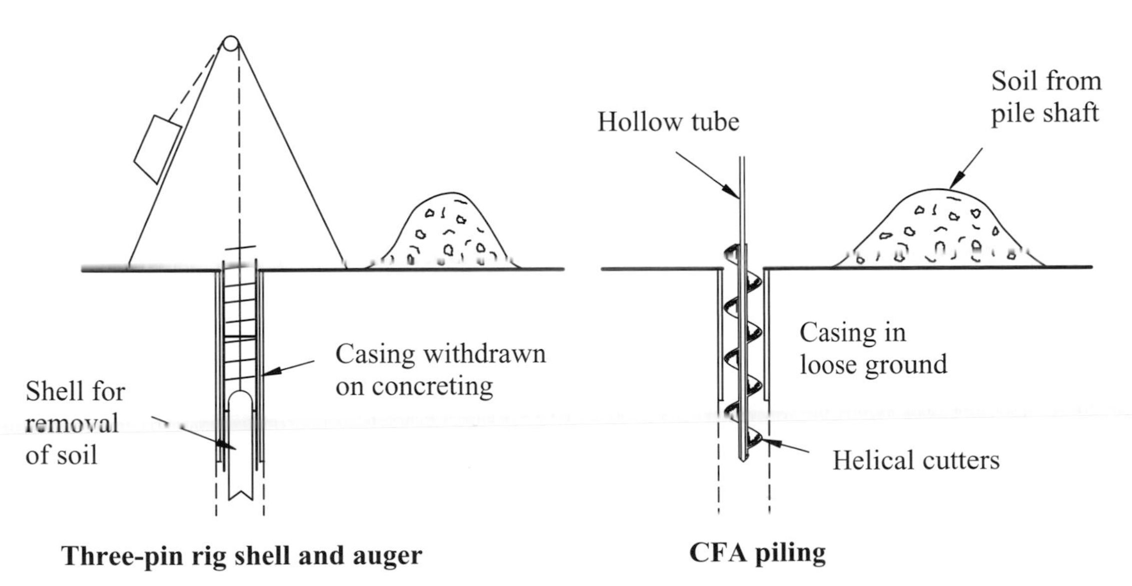

Pile design principles

Piles may be designed as end bearing piles or friction piles

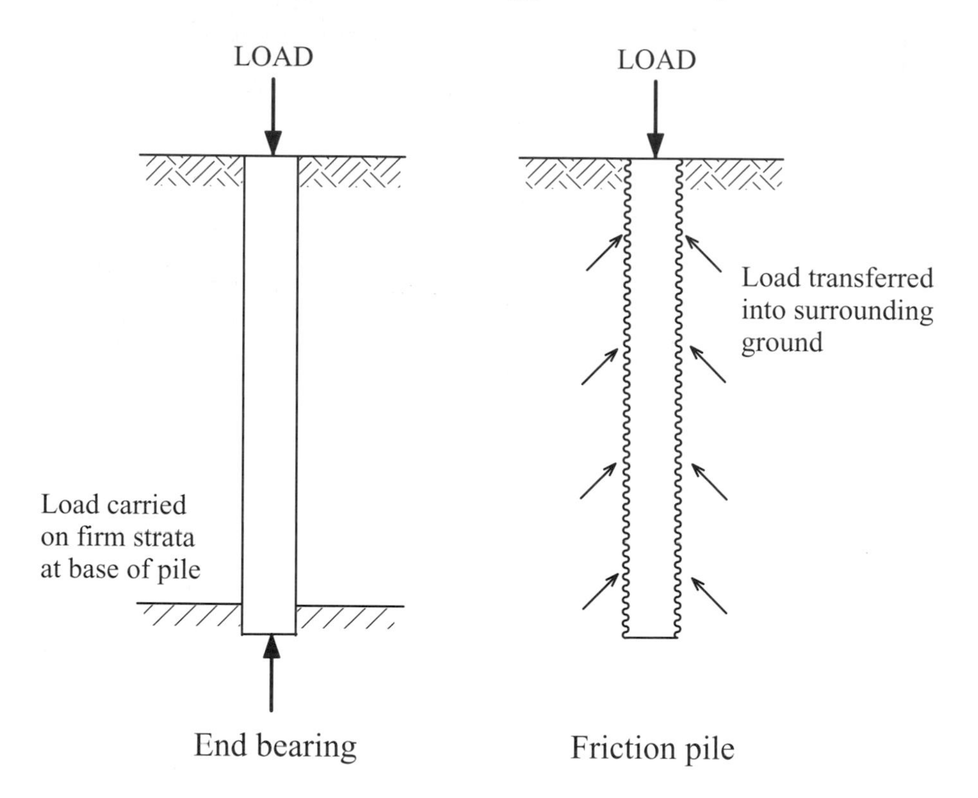

Displacement piles–driven pile techniques

During the driving of the pile shaft the soil is displaced to one side. Precast concrete piles, steel H piles and systems incorporating a shoe or plug fall into this category.

Replacement piles–bored piling techniques

During the driving or formation of the pile shaft the soil is removed and replaced with concrete. The main system in use is the continuous flight auger (CFA piling). These pile shafts are normally unlined.

A simple three–pin driving rig system may be used in difficult ground conditions. A sectional steel lining tube is inserted as the pile is bored and is withdrawn after placing the reinforcement and concrete.

4.5 Displacement piles–precast driven systems

Precast piles are normally square in section with chamferred edges. The bottom of the pile may be finished square or may contain a cast-iron pointed shoe.

The piles are lifted into the vertical position and driven into the ground using a drop hammer attached to the rig. Some of the piling equipment used by specialist contractors may appear to be a little antiquated, but this is the nature of the piling game. The piling gangs are made up of some of the 'hardest labourers' in the construction industry. Their work is strenuous and heavy (they often work in all weathers, often in restricted access conditions).

The piles are usually designed as end bearing and founded in firm strata. The piles are more suitable for projects where the driving length over the site area is at a constant driving depth. A thorough site investigation is therefore necessary to establish the ground profile and economic driving depth.

Precast piling is not recommended in built-up areas due to the noise created during the driving process. Piling close to existing buildings should be avoided where possible, so an open site is preferable.

Precast piles are manufactured in sections from 150mm to 350mm square. Standard pile lengths are 6–8m. The piles may be joined on site, based on a standard spigot and socket-pinned joint.

The piles are driven to a predetermined 'set' i.e. number of blows of the drop hammer to achieve a penetration depth of 20–25 mm.

Precast piles are relatively easy to handle and drive.

Specialist piling contractors include:

- Stent piling
- Bullivant piling
- West Pile
- M K Piling
- Keller Piling
- Abbey Plynford

Displacement piles

Precast piling and in-situ concrete ground beam foundation

Positioning the precast pile on the pile location

Driving the precast pile

Checking pile for 'set' (number of blows to drive the pile 25mm)

The project foundations incorporate precast piles supporting a reinforced concrete ring beam, brickwork to DPC and a beam-and-block ground floor slab.

The sequence of construction is illustrated.

Precast piling–displacement piles
Cropping the piles with a hydraulic pile cropper

Positioning pile cropper over pile

Group of piles to be cropped

Ground beam reinforcement in position

The sequence of work involves:

1) excavate the ground beam trench to expose the piles.
2) place the blinding concrete to the bottom of the trenches.
3) mark the cut-off levels on the piles and crop them.
4) fix steel reinforcement and concrete the ground beams.

Precast piling–displacement piles

Group of completed piles

Completed single pile

Completed reinforced ground beams ready for placing concrete

PC piles

The 200mm square precast concrete piles are 8–10m in length. Driving depth varies between 7 and 9m.

The piles are driven to a 'set' of ten blows per 25mm of driving. On excavation of the trenches, the piles are cropped with a hydraulic pile cropper.

Formwork and reinforcement are positioned to form an arrangement of ground beams. On concreting of the ground beams, brickwork is constructed up to the floor slab level.

Brickwork to DPC level and fill to floor areas

Beam and block floor forming ground floor slab

Precast piling using a mini pile driver

The illustration shows the use of a tracked mini pile driver used in conjunction with precast concrete piling. Short driven pre-cast piles up to 5m in length are being installed for a single-storey office project.

Displacement piles

Mini-piling methods - 190mm diameter steel tube with sealed end

Plan of base showing six mini piles in position

Rotary boring rig for inserting lining sleeve to the top 2m of the pile shaft (see sketch on page 110).

Equipment for rotary boring and vibrating equipment for inserting the 190mm diameter piles.

Sequence of installing mini-piles driven (displacement piles)

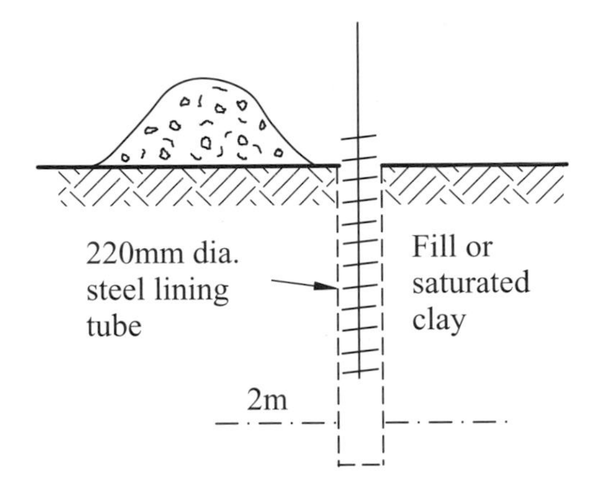

Stage 1 –Insert 220mm dia. lining tube

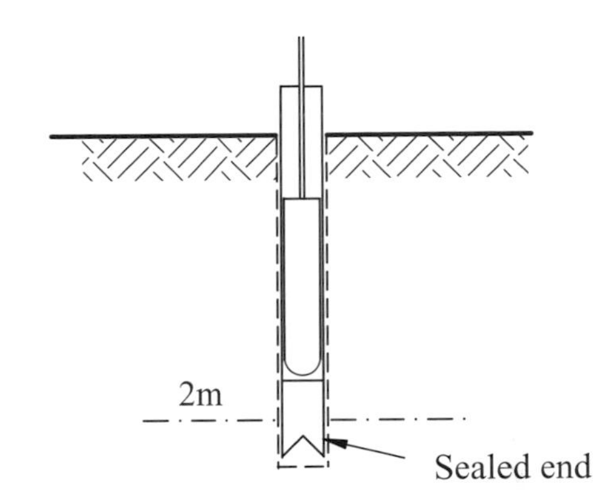

Stage 2 –Insert 190mm dia. steel pile with sealed end.

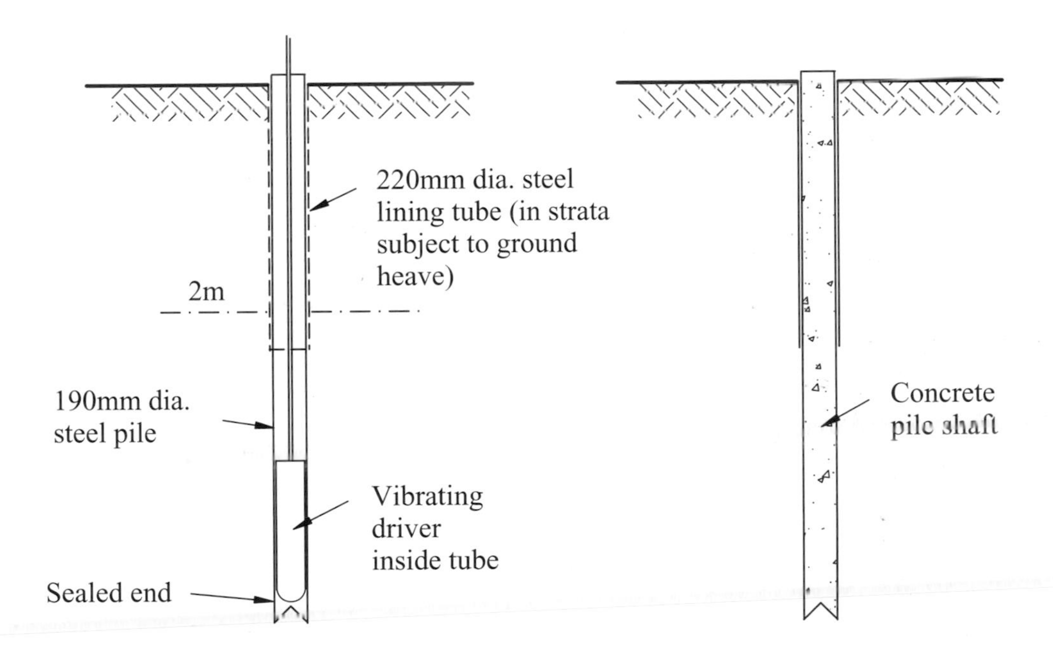

Stage 3 –Drive 190mm dia. steel pile to design depth.

Stage 4 –Concrete pile shaft insert central steel reinforcement bar.

Sequence of installing mini-piles

220mm dia. outer sleve with 190mm dia. pile driven inside

Driving 190mm dia. pile with vibrating tool–pile sections approx 2m in length

Welding the spigot and socket joints between pile sections

Vibrating 190mm dia. piles. On completion, pile shaft is concreted

4.6 Replacement piles–continuous flight augered bored piling system

CFA piling

CFA piling is the most successful form of replacement piling method used in construction.

The equipment/plant required to undertake the process is shown above:

- piling rig.
- excavator and dumper to remove spoil from the pile shaft
- wet storage hoppers and pump for placing the concrete.

The piling gang consists of five operatives (one machine driver, three labouers, one concrete operator).

Ten to fifteen 450mm diameter piles can be formed per day (up to a depth of 18 metres per pile).

This represents approximatly 1.5 piles per hour.

CFA piling

The piles are formed using a hollow stem auger. A helical cutting edge is continuous along the full height of the auger. A hollow tube runs down the centre of the shaft and this is used for pumping concrete into the pileshaft as the auger is withdrawn.

The discharging of soil from the auger is shown around the top of the pile at ground level. The spoil is removed to a tip, on or off site' using an excavator and a dumper.

Pumped concrete from the wet storage hoppers is fed into the base of the rig and pumped into the pile shaft.

The overall process of forming the pile shaft is relatively simple and effective.

CFA piling system (400mm diameter piles)–bored pile system

CFA piling–work sequence

Set up piling rig over pile position.

- Bore pile shaft and remove spoil by excavator.
- Concrete the pile shaft as the auger is withdrawn using a wet hopper and ready-mixed concrete supply.
- Lower steel reinforcement into position.

Auger disharges soil at side of rig

Placing steel cage reinforcement in 13m deep piles

4.7 Ground treatment
Vibro techniques–displacement piling technique

The vibro-replacement process was originally developed by Kellar Foundations in the early 1950s. This system of treating filled ground is the most popular method of ground improvement for low-rise buildings. The process is considered economical, effective and a well-proven method (Warren,1996). The area of compaction need only be the area directly under the foundation.

The vibro-replacement process (stone columns)

The diagram illustrates the principles of the vibro-replacement system for the installation of stone columns using both the 'wet technique' and 'dry techniques'.

The wet technique introduces water jetting when forming the 450–800mm diameter hole prior to the stone being placed from ground level. The dry technique simply uses the vibrator to form the hole, which is then fed with dry inert clean stone from ground level.

Process of forming stone columns: 'drytechnique'

The sequence of forming the stone columns is:

Stage A–Position the vibrator over the pile location (this is indicated by positioning steel pins into the ground at each pile location).

Stage B–Upward and downward movements of the vibrator allow the hole to be formed to the required depth.

Stage C–Clean angular stone (40–75mm) is introduced into the hole from ground level (placed by shovel or by an excavator bucket). Further vibration takes place in order to force the stone into the surrounding ground.

Stage D–Further vibration takes place as additional stone is placed in the hole as compaction takes place up to ground level.

On average 20–30 stone columns may be formed per day.

By arranging the centres of the compaction points, the stone columns overlap. This enables full compaction over the area of the foundation to be achieved. The layout pattern of the stone columns varies according to the foundation arrangement and loadings. Penetration depths vary according to the data available from the site investigation report. Normal stone column depths are 3–5m.

A strip foundation pattern is shown on page 119 for a narrow foundation and an alternative wide strip foundation. The pattern used under a raft or ground-floor slab area is also illustrated.

Vibro-replacement techniques formation of stone columns

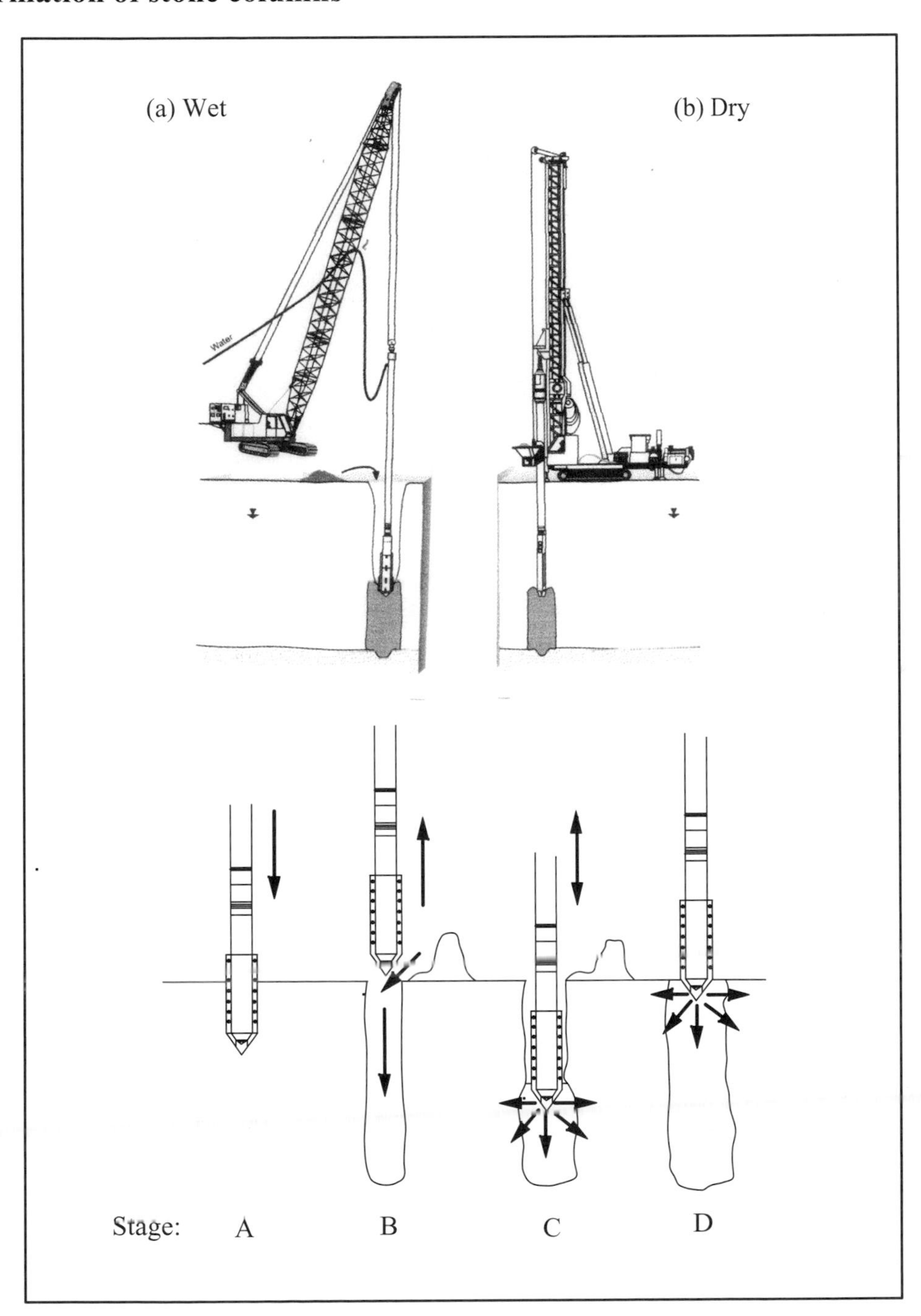

Vibro-replacement

General view of pile rig–requires a separate generator to power the vibrator

Setting out the pile locations on 1.2 metre grid layout

Vibro-replacement

Vibro-replacement is a relatively simple and speedy process. Twenty to thirty stone columns of approximately 3m depth may be completed in a day. Only two items of plant are used in the installation process: the pile rig (and generator) and a side discharge skip loader for handling the stone aggregate to aid the filling of the stone columns.

Note: Stone columns are formed with the vibrator in the vertical position.

Equipment required for process, i.e. piling rig plus power shovel with inclined tipping facility

Rig in position to commence drilling
Vibrator positioned over peg in ground

Vibration in progress, upwards and downwards movement

Placing stone aggregate in place as further vibration takes place

Vibrator at full depth

Completed strip foundation on completion of concreting

Vibro-replacement

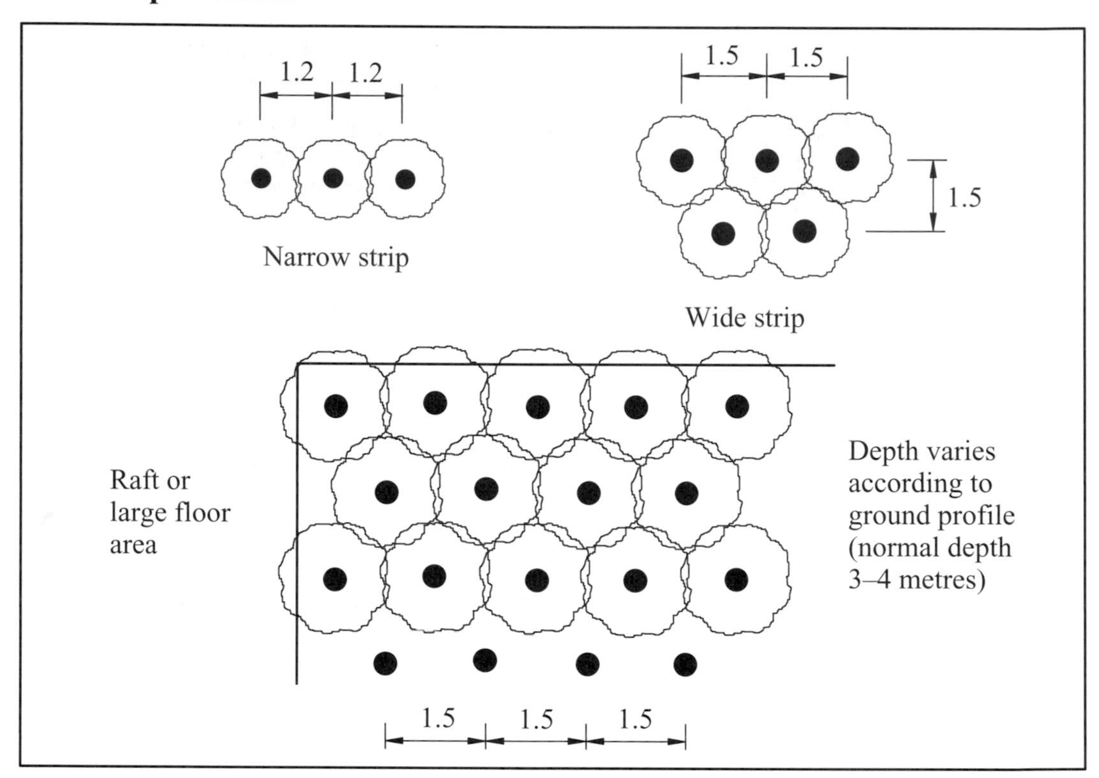

Alternative patterns of treatment

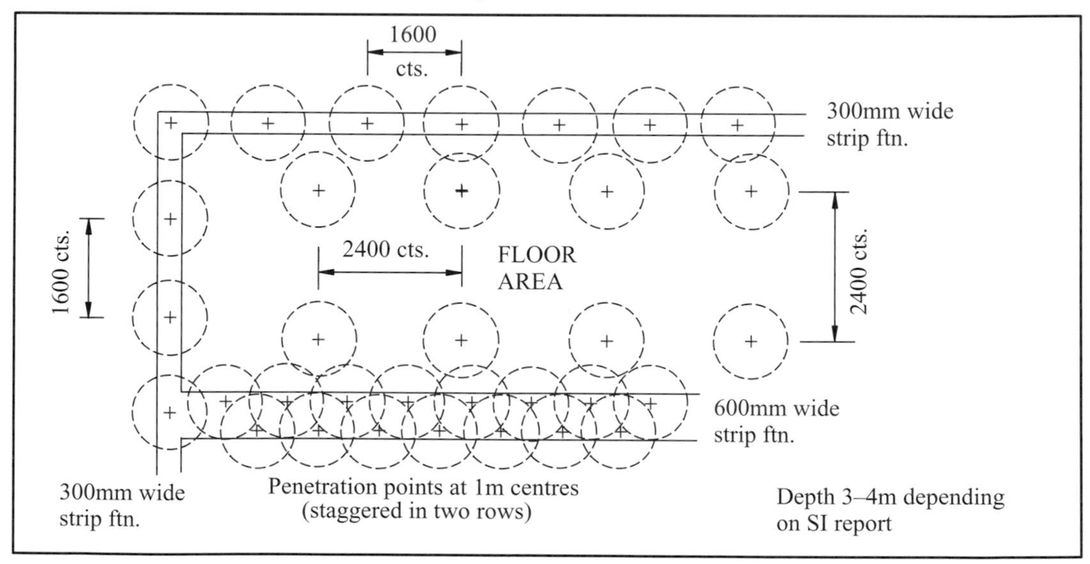

Layout pattern used on project

CHAPTER FIVE

EARTH SUPPORT AND BASEMENTS

5.0 Overview

Earth support principles and applications are outlined for service trenches, drainage and manhole excavations. Images are shown of trench and manhole boxes.

The principles of earth support using sheet piles is indicated together with applications to deep excavations.

The use of H piles used in conjunction with precast panels, timber sleepers and steel sheets are illustrated for with reference to construction methods. The use of battered excavations is shown in a number of site situations.

A case study is presented on the construction of a bored piled secant wall to a basement project, to form an underground car park.

Groundwater control using well points is illustrated by a project in Italy.

The various methods of constructing basements are discussed and case studies are presented on the principles and applications of precast concrete basements on two UK projects.

Once again, the principles and applications of brick and reinforced concrete basements are covered in a wide range of books published by UK authors.

What is happening here?

5.1 Earth support considerations

The degree of earth support required for trench and ground excavation is directly related to the ground conditions, depth of excavation and location of proposed work.

Earth support methods

Service trenches	Timber support/poling boards and struts Steel trench sheets
Drainage trenches and manholes	Trench sheets Drag boxes or trench boxes Manhole boxes Sheet piles
Deep excavations and basements	Steel sheet piles Bored pile / secant piled walls H pile support methods
General excavations	Battered excavations in good ground. Integration of dewatering systems in high water table situations.

Factors influencing earth support

For any excavation over, say, 2 metres deep some form of ground exploration needs to be carried out. This may vary from an excavated trial pit to a full site investigation.

Factors which need to be considered include:

- ground conditions, from information available
- depth of excavation
- groundwater table levels
- location–space restrictions around working area
- degree of specialisation likely to be involved, i.e. sheet piling, dewatering etc.
- proximity of other buildings and roads

The preparation of a method statement and hazard assessment should always precede the commencement of excavation works.

5.2 Earth support to service trenches and drainage

Service trench excavations are normally at depths up to 1.0 metre. Simple earth support is usually provided by inserting short timber poling boards, waling and cross-trench props (usually Acrow prop types). For deeper excavations in constructions with manholes or drain connections, trench sheets may be used.

Trench boxes (known as drag boxes) in use for a 3m-deep excavation

Hydraulic manhole box

When working in deep excavations where the ground is liable to collapse, trench boxes (or drag boxes) may be used.

Manhole boxes may also be used. These steel-faced boxes are held apart by hydraulic struts. Heavy plant is required to handle them into position.

Trench box being handled with an excavator

Hydraulic trench sheets in manhole box

5.3 Sheet piling support principles

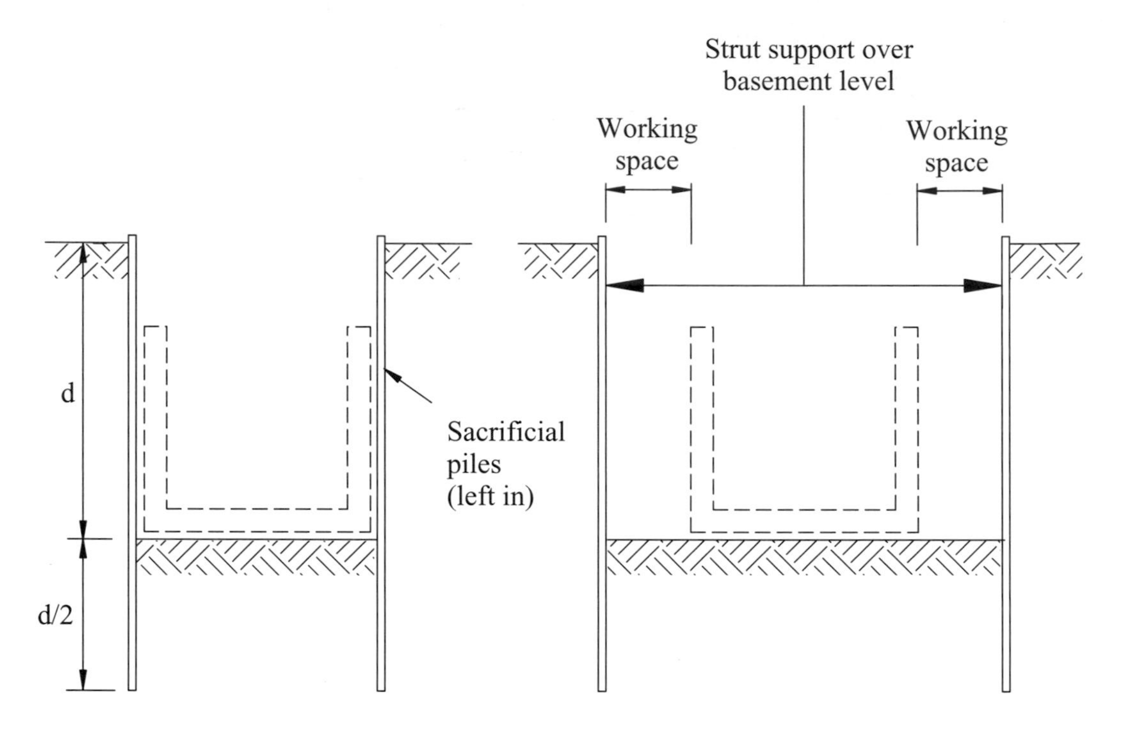

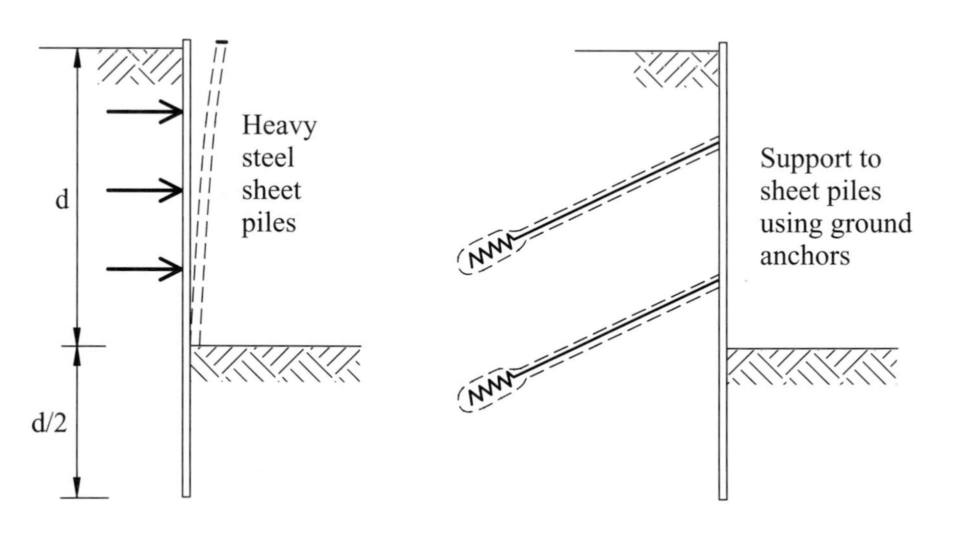

Free earth support principles

Earth support in poor ground conditions

Sheet piled deep excavation
Sheet piling provides a cut-off to ground water entry

Sheet piling to an underpass

Sheet piling to a deep excavation using steel walings and struts

Handrail located around the excavation

Water entering the excavation may be pumped from sumps in base of excavation

View of sheet pile profile

A series of images for earth support to a bridge pier basc– excavation approx. 4 metres wide x 12 metres long x 5 metres deep.

The excavation has been piled with Frodingham steel piles. RSJ sections have been used for the walings in conjunction with steel and hydraulic struts.

Excavation work has been undertaken with grab attachment working between the sheet piles.

5.4 Use of steel H piles with precast infill panels

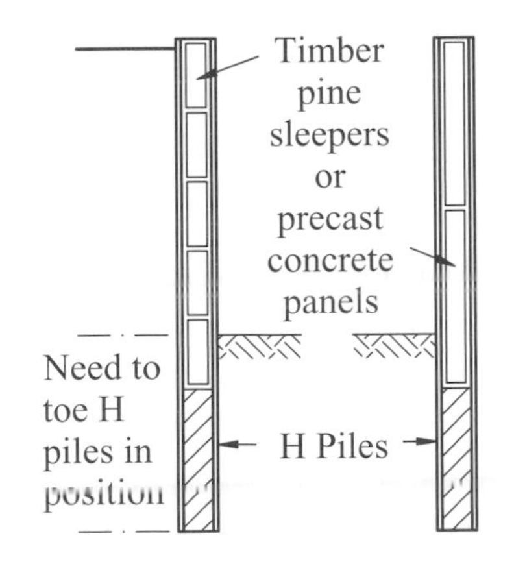

The process involves inserting steel H piles in a vertical position around the excavation. As excavation takes place, the timber sleepers or precast concrete panels are positioned between the H piles. This acts as permanent support to the excavation and as a surface against which to construct the basement wall.

Temporary propping to H piles is shown prior to constructing the floor bay adjacent to the wall.

Refer to applications in Germany shown on page 129.

SECTION

PLAN

Use of steel H piles incorporating steel plate supports

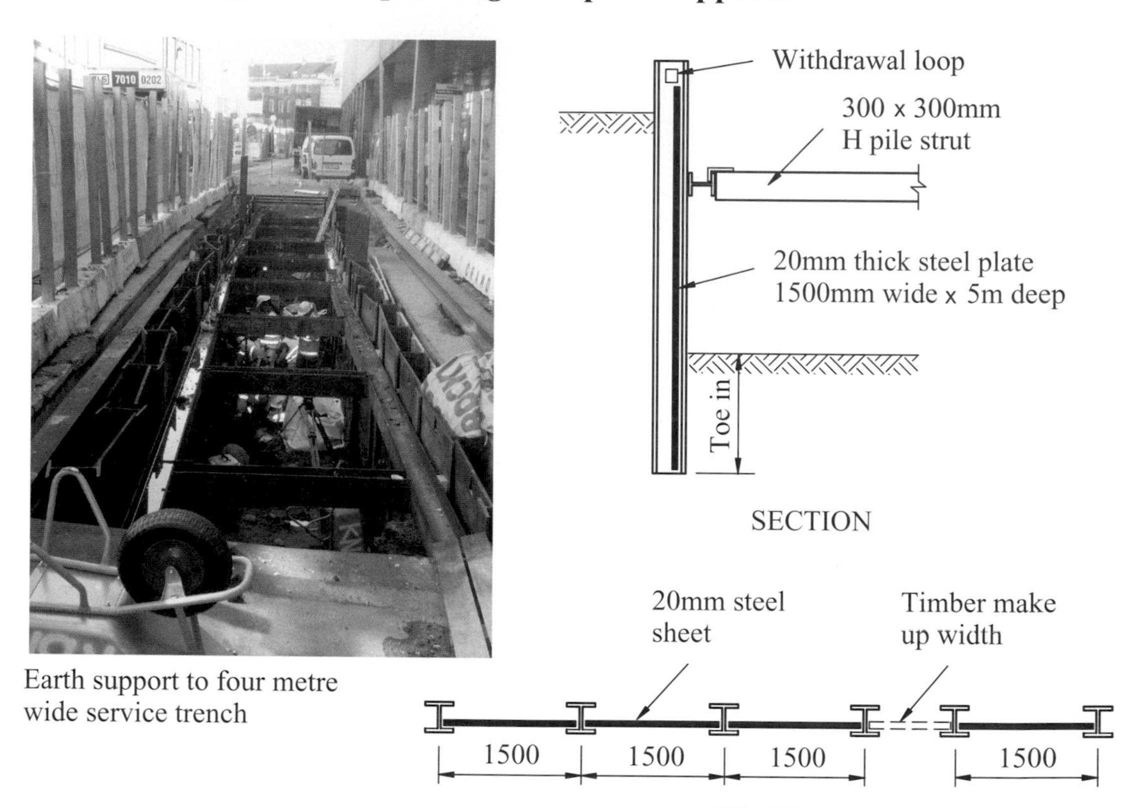

Earth support to four metre wide service trench

The technique consists of positioning 20mm thick steel vertical plates between H piles inserted at 1500mm centres.

Elevation of H piles with vertical steel sheets in position

View of corner arrangement of H piles and steel plates. Note the facility to enable easy removal of steel plates on completion of the works

Four metre wide excavation for service trench in major road. H pile support to both sides of main excavation

Applications of earth support techniques in Germany

Use of H piles with horizontal timber sleepers fixed between
Note ground anchors used to secure H piles in position

Use of H piles with vertical precast concrete panels
to support deep excavation–an interesting approach.

5.5 Earth support–battered excavations

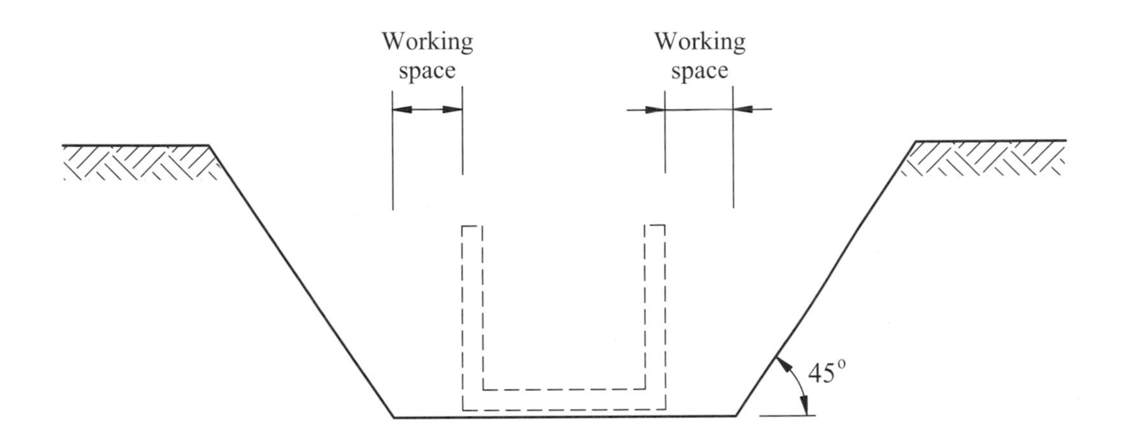

No groundwater problems
Battered excavations

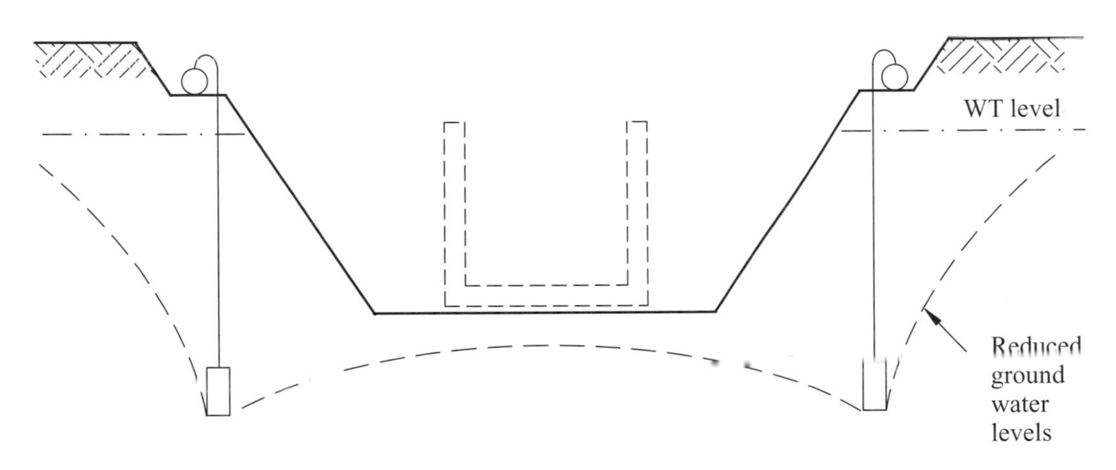

Groundwater problems
in sandy ground
Dewatering/well-pointing

(open site–no boundary problems)

Battered excavations to a basement and general site situations

A collection of images of battered excavations in a variety of ground conditions varying from bolder clay to compact sandy gravel.
Two-metre battered excavation to a large foundation base

The 45° angle of repose proposed totally ignored here

Battered excavation to a basement

The use of battered excavations on large open sites in good ground is common practice. The angle of repose varies according to ground conditions and stratification of the soil (the way soil layers of different material lie one on top of one another).

5.6 Sequence of forming a secant bored pile wall

Construction of guide walls

Part excavated basement illustrating exposed pile wall
and capping beam (quantity of excavation 8000 c.m.)

General view of completed guide walls one metre deep
Note the formation of pile profile between guide walls

Sequence of forming a secant bored pile wall

1. Construct guide walls–Average 1m deep

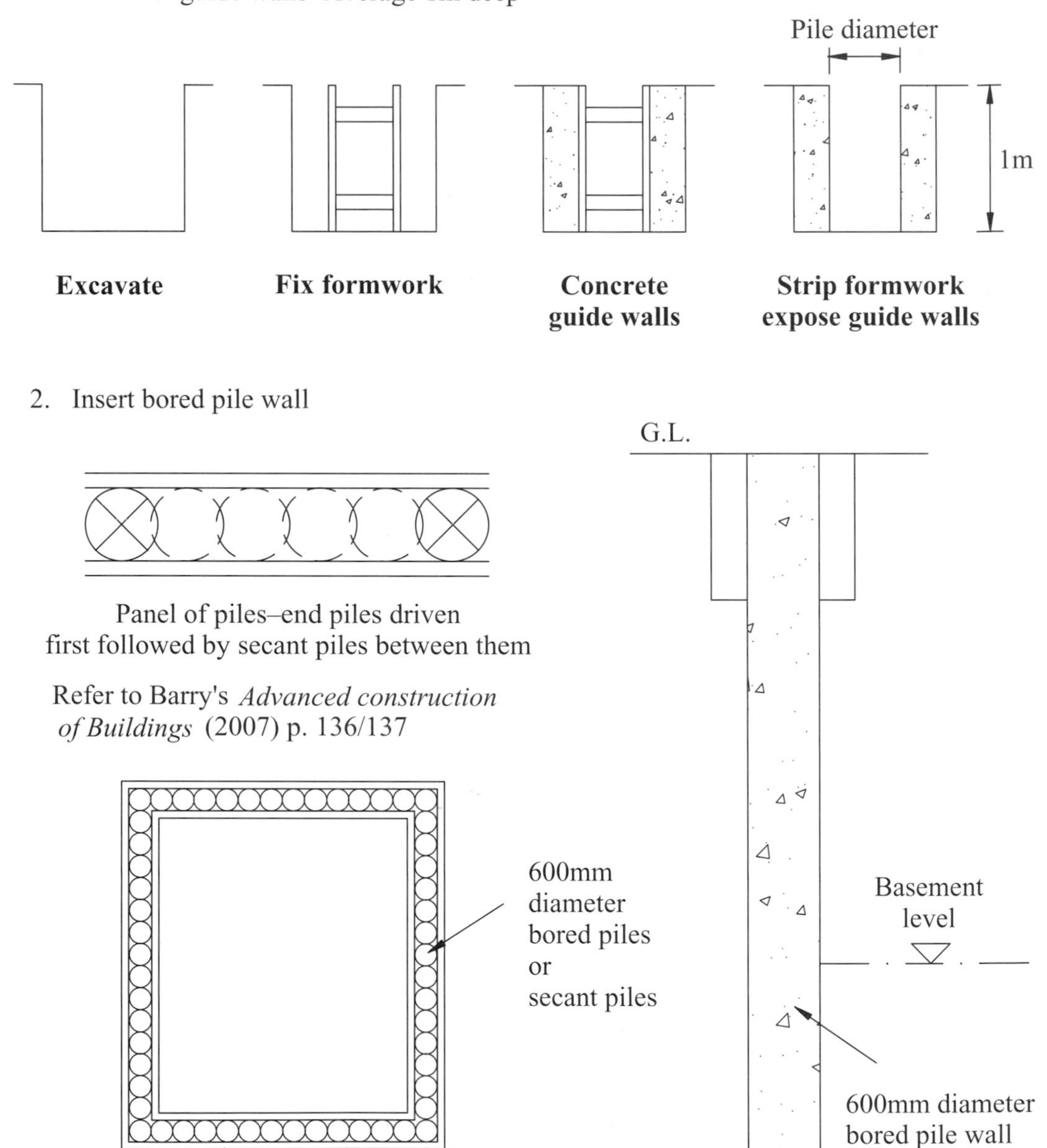

Commencement of boring.
CFA piling system

Piling rig positioned over
guide walls

View of pile panel.
End piles driven first

Breaking out guide walls
after completion of piling

The basement size was approximately to 50m x 38m on plan. A vehicular access ramp is provided into the basement area.

On completion of the basement excavation a 300mm thick-stone piling mat was laid over the basement area. This was to accommodate the piling rig to sink 200 CFA piles to support the basement floor slab.

3. Break out guide walls and form capping beam

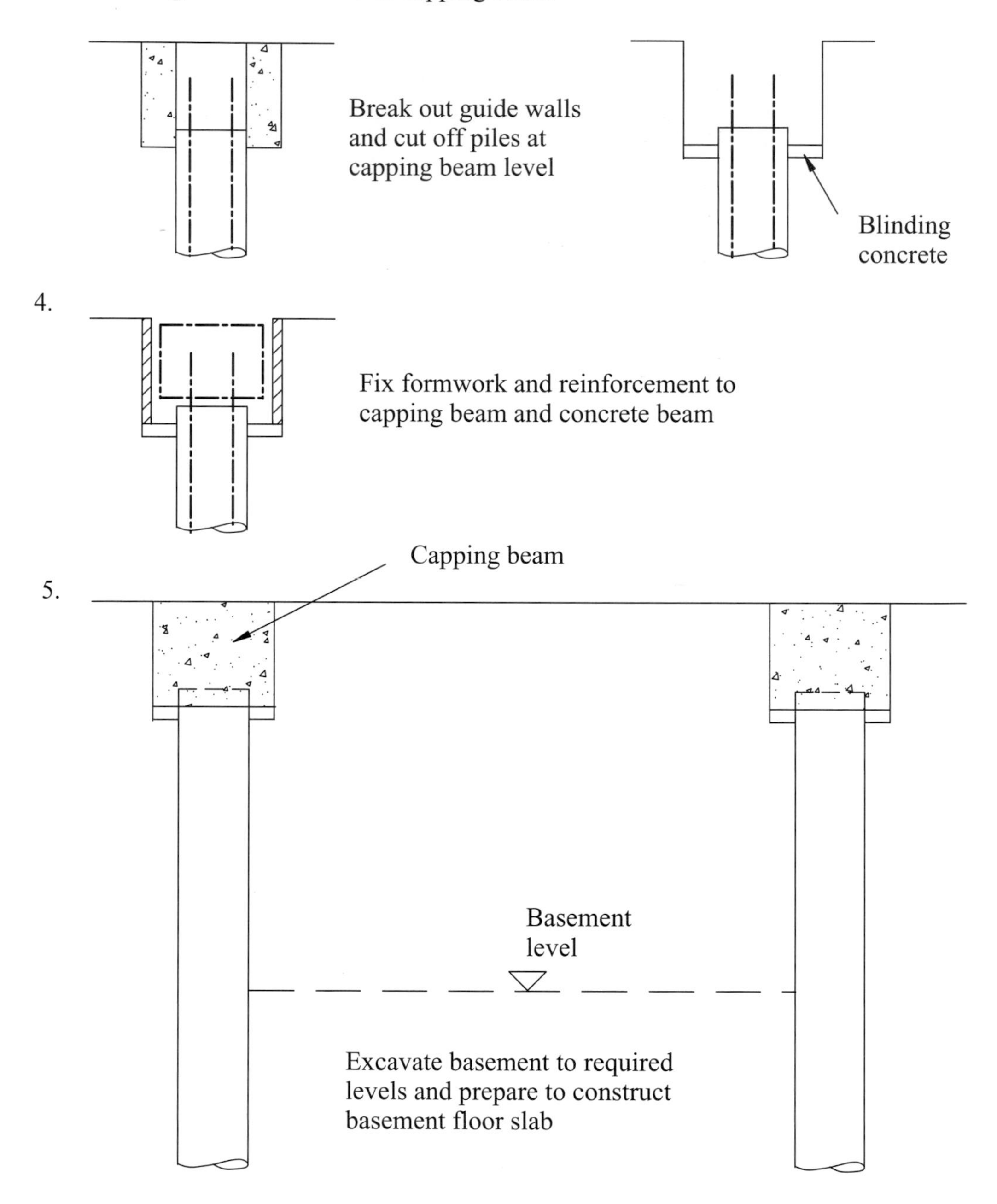

Fixing capping beam reinforcement and formwork

Competed capping beam

Ground view of capping beam and formwork

Piles after cropping

Formation of bored piled wall

The stages involved in the construction of the bored piled wall are as follows:

- set out basement
- construct 1m-deep concrete guide walls
- undertake CFA piling to the perimeter of the building pile–depth approx. 10m
- break out concrete guide walls and cap off piles to the level of the underside of the capping beam
- construct reinforcement/formwork and concrete to capping beam
- excavate basement

5.7 Groundwater control considerations

Various methods may be considered for dealing with groundwater encountered in excavations. The method selected is dependent upon ground conditions, water table levels and site location.

The following methods may be considered:

- sheet piling or a bored pile wall constructed in order to cut water off from entering the excavation
- dewatering (or well pointing) to lower the groundwater level while the work is carried out
- Pumping– from sumps or wells. Pumping from sumps may take place prior to or during the excavation process, and submersible pumps may be used to control water levels once the building is in use

Dewatering may be used in an open site location where the soil is granular. A battered excavation may be formed, and the dewatering system used to lower the water table prior to the commencement of work.

Dewatering may also be used inside a sheet piled excavation to reduce the water table locally. In wide excavations a combination of sheet piling and dewatering may be specified (as in the following case study).

5.8 Dewatering case study

PROJECT: CONSTRUCTION OF A NEW SWIMMING POOL COMPLEX IN ITALY

Project details:

Ground conditions: course-grained sand, free-draining strata

Water table 1 metre below original ground level

Excavation depth–max 3 metres

Pool size approximately 25m x 20m.

Earth support to pool area to be formed in Frodingham heavy sheet piling

Area outside reduced to a depth of 1 metre prior to piling and installing dewatering system

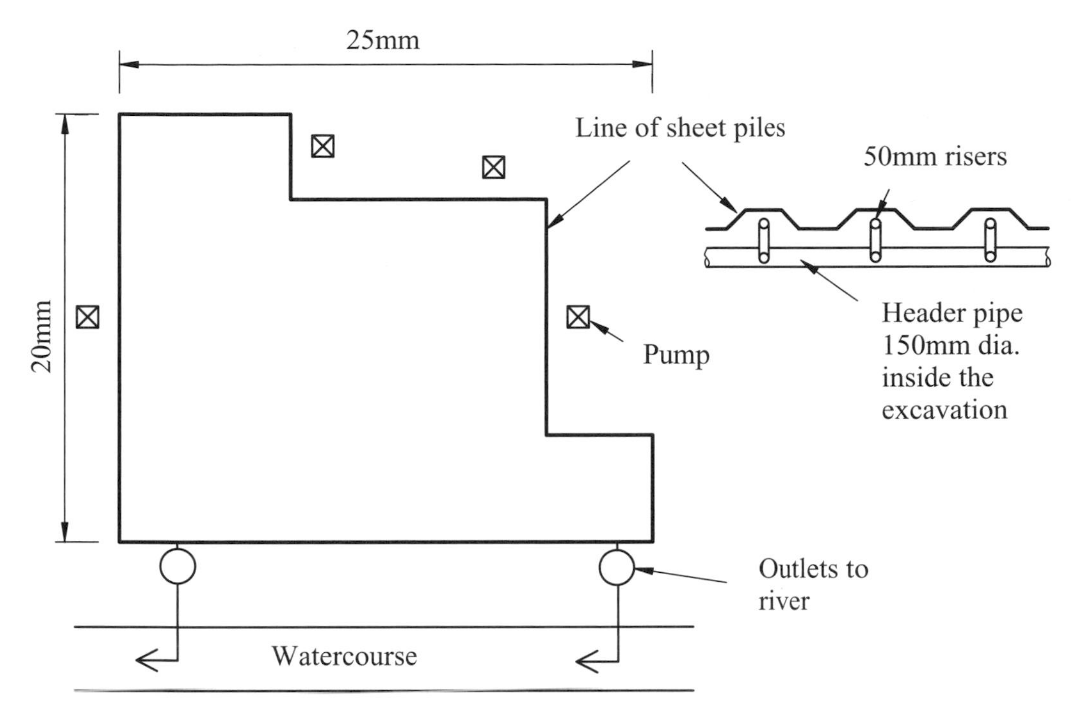

Plan of excavation to pool area

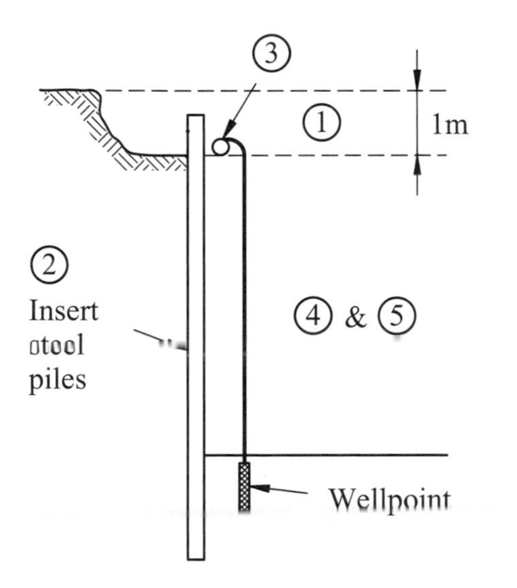

STAGES OF WORK

1) Excavate oversite area 1m deep.

2) Insert steel sheet piles to perimeter of excavation 6 metres deep.

3) Locate well point pumps, position 150mm dia. header pipes and insert well points to a depth of 4/5 metres.

4) Commence pumping for 3/4 days to reduce water table level inside pool area to be excavated.

5) Commence excavation work to main pool area.

Note: header pipe slung on steel straps to hold it in position during the excavation works.

Layout of well point header pipe around excavation

General view of site showing the layout of the well point header pipe and sheet piles. The ground on the site area has been excavated to a depth of 1 metre prior to piling and inserting the well points.

Connections from header pipe to well point riser

The 150mm diameter header pipe is located around the excavation perimeter. The 50mm flexible hose connections can be seen connected to the well point risers. The red flexible pipes are carrying pumped water to the outfall.

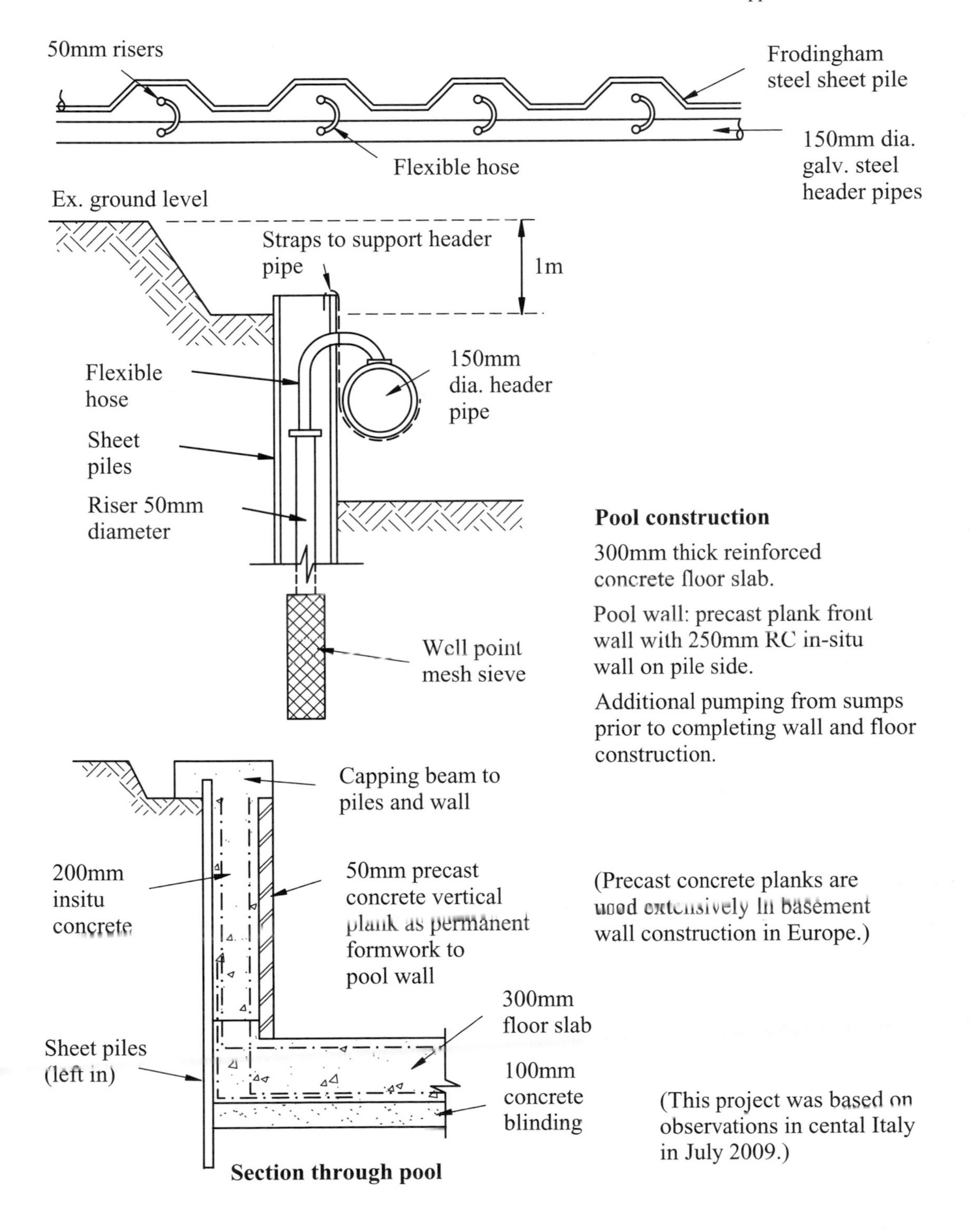

Section through pool

Typical 150mm dia. pump

Four 150mm diameter pumps are located around the excavation perimeter. The system has two outfalls into a watercourse adjacent to the site.

View of header pipe and risers located between the sheet piles

Six-metre-long steel Frodingham piles have been driven around the perimeter of the site.

As excavation precedes to a depth of approx. 3 metres, the well point risers can be readily seen in the web of the sheet piles.

Pumping outfall to adjacent watercourse

It is envisaged that further well pointing may prove to be necessary to the central area of the excavation.

Additional pumping from sumps adjacent to the sheet piles will also be necessary as the pool base slab is constructed.

Excavation work in progress

The loose granular course-grained sand is easy to excavate. The excavated material is being deposited on site for filling low areas adjacent to proposed building work.

5.9 Basement construction methods

General considerations basements may be constructed in the following ways:

In-situ concrete basement

In-situ concrete floor and walls with either an in-situ or precast concrete roof (i.e. to form the ground floor slab). High quality concrete may be specified and a water bar (either metal or PVC) may be incorporated in wall and floor joints. Full height formwork to the wall pours is normally specified as this reduces the number of wall joints. If the work is properly specified and supervised then no other waterproofing is required. Consider the number of elevated sewage tanks that are constructed above ground and do not leak.

Traditional brick basements

Basements may be constructed of brickwork on a concrete base slab–again a precast or in-situ concrete roof may be incorporated in the construction.

Brick basements require waterproofing, often to the floor and walls. This may be provided in the form of horizontal and vertical asphalt layers or by proven products such as Bituthene ® or other sheet applied materials. Care is required around entry points for pipes and service ducts.

Precast concrete basements

The use of precast concrete basements has been developed on the continent over the past few years and is now becoming an economical alternative in the UK. The systems developed use either a single wall construction or a double wall, infilled with concrete.
The 50mm precast wall slabs are used as permanent formwork to speed up the construction process. Two case studies are illustrated within the text.

Basement construction methods

In-situ concrete

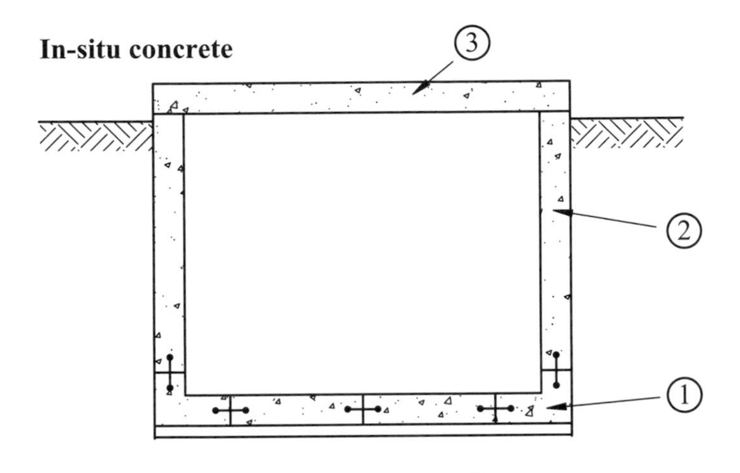

Sequence of work

1) Construct floor slab
2) Construct walls, reinforcement, formwork, concrete
3) Construct roof slab in-situ or precast

Traditional brick

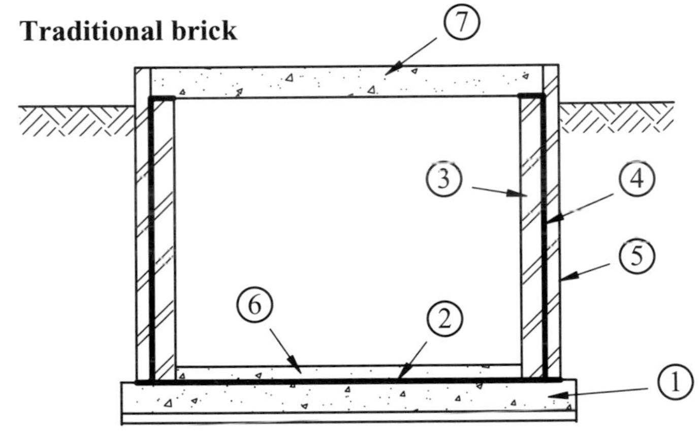

1) Construct floor slab
2) Asphalt floor surface
3) Construct inner brick wall
4) Asphalt wall
5) Construct protective brickwork skin
6) Construct inner floor
7) Construct roof slab

Precast concrete

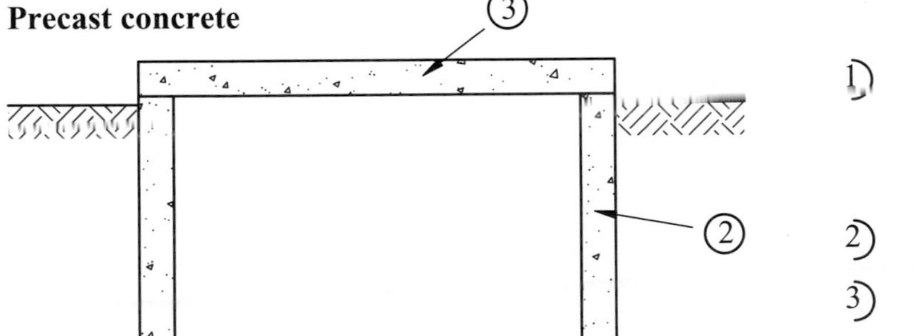

1) Construct floor slab, reinforcement, formwork, concrete
2) Erect precast walls
3) Construct roof slab–precast or in-situ

5.10 Precast concrete basements

A basement may be constructed with a single-skin precast concrete wall, or two pre-cast walls with the cavity filled with concrete. These systems originated on the continent and are widely used in France and Germany. A number of German contractors now offer the service in the UK. (Glatthaar, Huf Basements and others - see internet web site references).

Thermosafe basement

The construction of a Thermosafe basement consists of:

- a 250mm thick in-situ concrete floor slab. Starter bars protrude from the slab and are accommodated within the wall cavity. These will form the structural link between the walls and foundation once the concrete is poured
- the wall panels, which are precast off site, are craned into position over the starter bars. The panels are temporarily supported with props ready to receive the semicast roof panels on top. The outer wall panel consists of a 50mm precast panel with 70mm of insulation fixed on the inner face. The overall wall thickness is in the order of 350mm
- the 50mm-thick precast concrete deck slabs are now positioned and temporarily propped at 1500mm centres
- the wall cavity is now filled with ready-mixed concrete. On completion of the wall pour, the 200mm in-situ concrete slab is poured to form the ground floor area of the building.

A section through the basement wall construction and roof slab is shown on page 148.

All external horizontal and vertical joints are filled with a waterproof flexible grout, and a bitumin-based tanking is applied to the external walls. The external tanking is protected by an insulating sheet prior to backfilling the area.

Other precast systems

Consist of using two 50mm thick precast concrete walls filled with 170mm of in-situ concrete. The roof slab consists of a propped 50mm concrete deck slab supporting a 120mm thick in-situ concrete slab. Externally the basement wall is further protected with a DPM membrane and a 75mm thick bead insulation slabs.

Construction sequence–Precast basement

Erection of basement precast walls

Basement slab complete fixing first wall panel

Erect internal basement walls

Patent precast basement system

Precast concrete basements are a popular form of construction in European countries. These systems are now available in the UK by specialist contractors often linked to German partners.

See: Basement information centre web site reference

Erect wall panels

Positioning precast roof units
Wide slab deck units

Precast basement detail

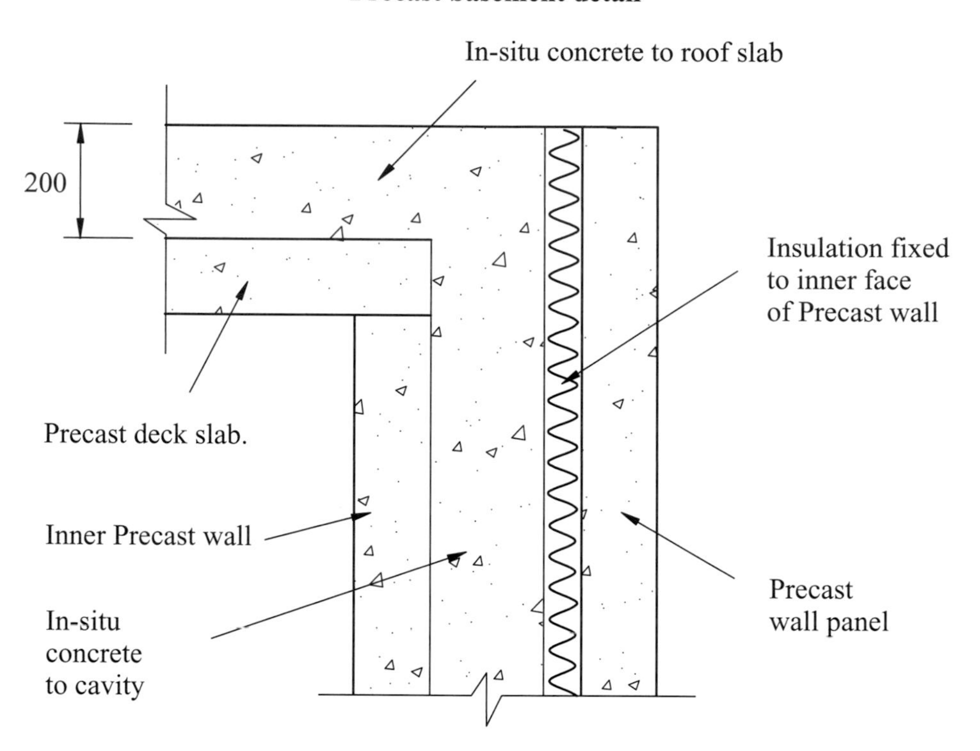

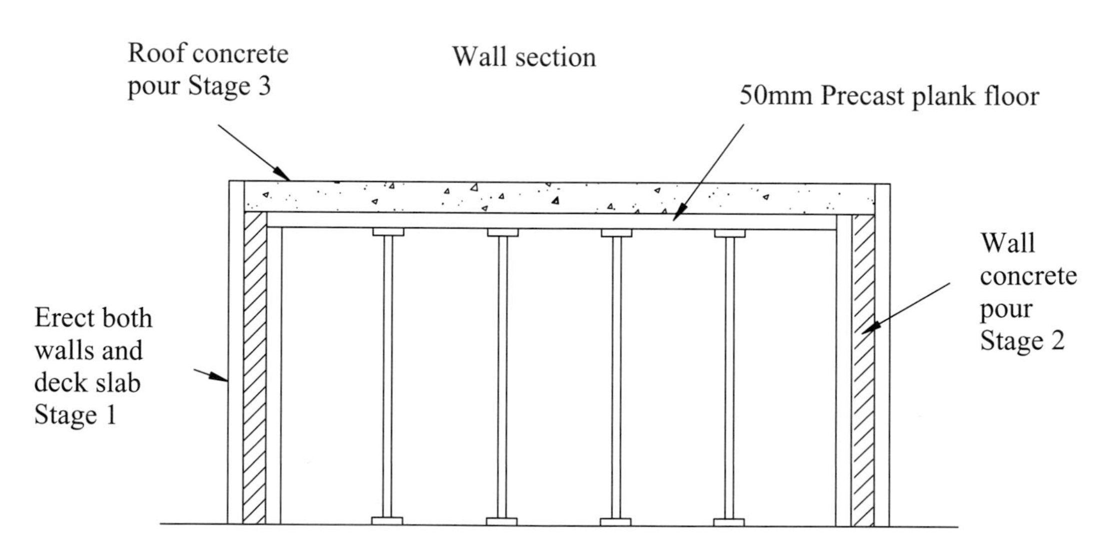

5.11 Precast underground storage tank

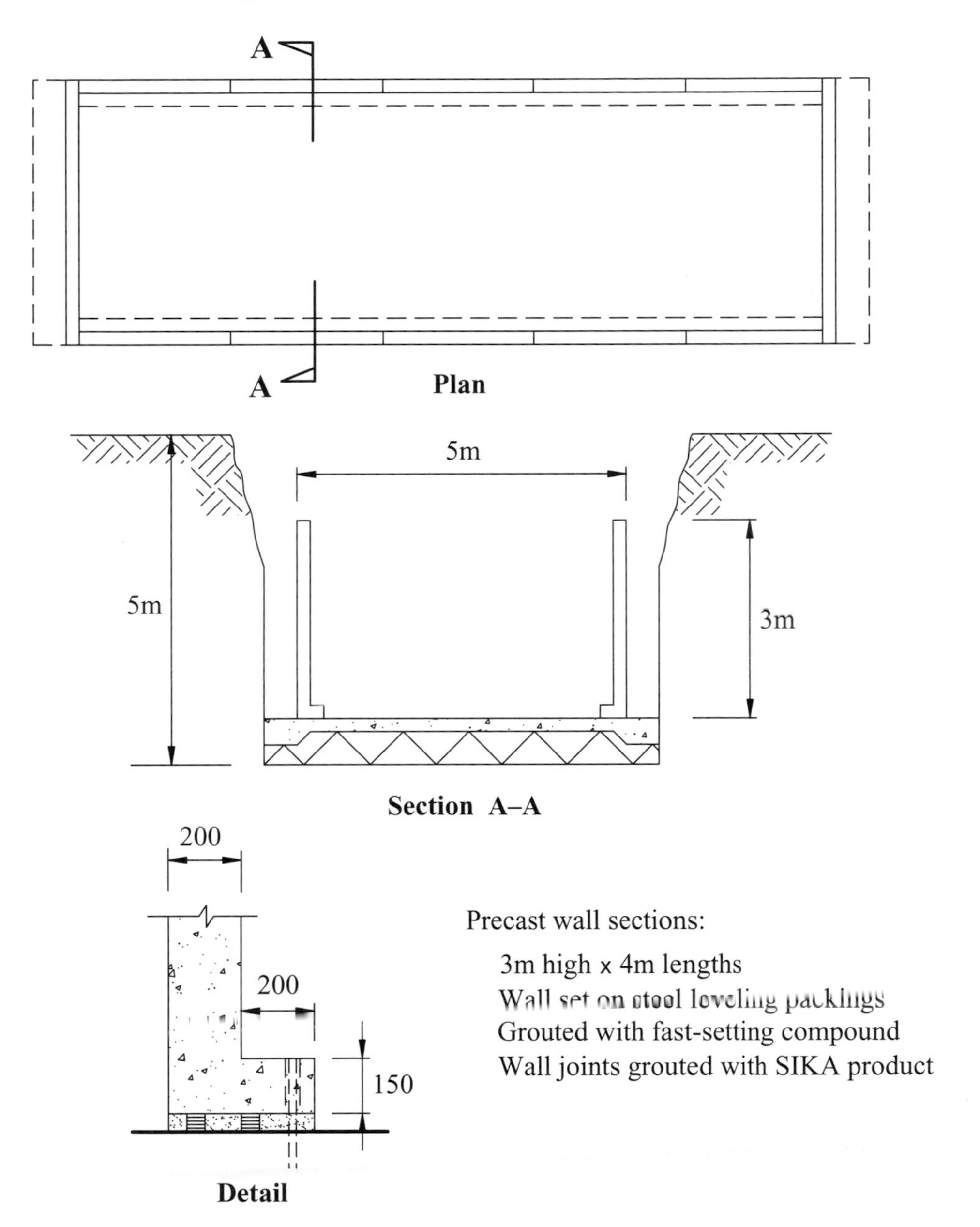

Work sequence:

1) Excavate pit
2) Stone up base
3) Fix base reinforcement
4) Concrete floor slab
5) Fix precast walls and deck roof

Construction sequence

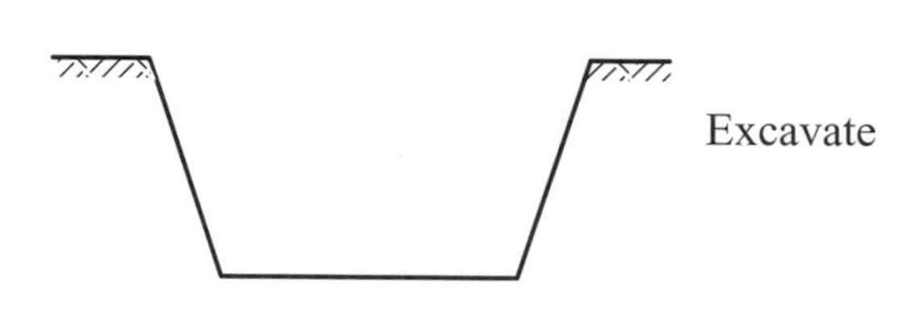

Stage 1

Excavate pit
8.0 x 6.0 x 25 = 1180 c.m.

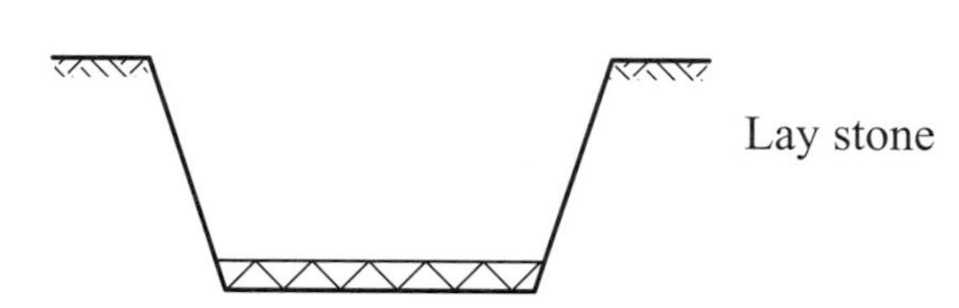

Stage 2

Lay compacted stone base of pit average 300mm thick = 50 c.m.

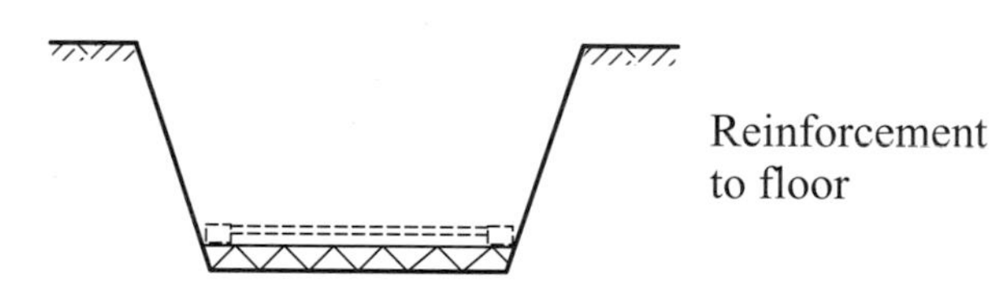

Stage 3

Fix reinforcement to perimeter edge beam and fabric to base

Note: the slab is to be cast back to the excavation sides –hence no edge formwork used

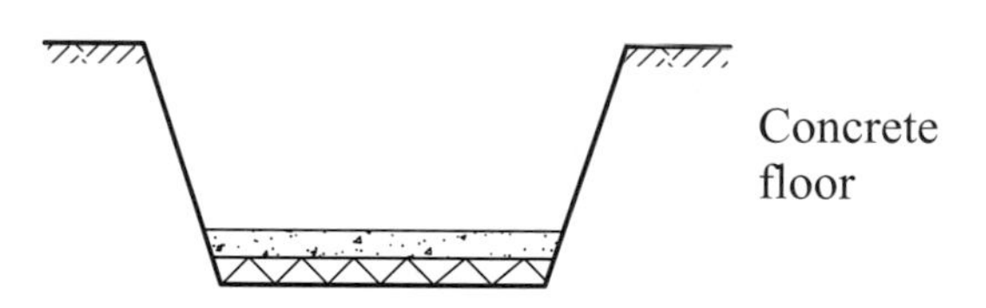

Stage 4

Concrete the slab base–300mm thick slab and edge beams Approx. 60 c.m.

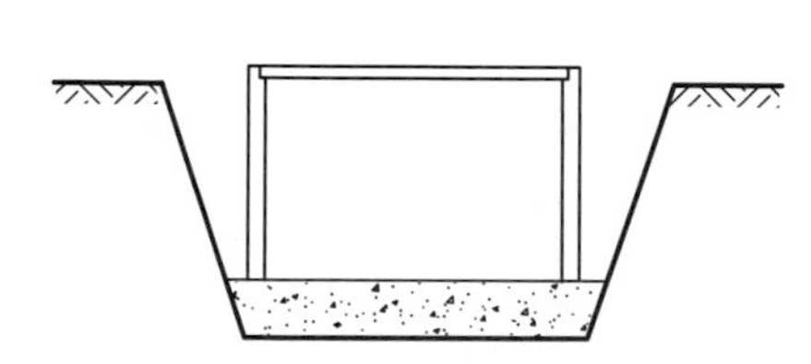

Stage 5

Erect precast concrete walls and roof beams (wide slab precast floor units)

STAGES OF WORK

The sequence of construction is illustrated on page 152.

General view of excavation to pit
Excavate using large hydraulic back actor load, 10 c.m.
lorries and remove to tip off site–duration approx. 4 days.

Formation of ramp for access to base slab level

Stone up base slab and commence fixing of reinforcement to edge thickening–duration 4 days

Concrete pump located on road adjacent to the pit, slab base–60 c.m., pour placed in 4 hours, four labourers in placing/finishing gang

Placing concrete in slab–note the very firm/stiff clay strata

Levelling to surface of concrete using a rubber panel float

Erect end precast panel

Delivery of precast panels on A frames

Erect side walls

Progress at the end of day 1

Day 1

Erect 200mm thick precast walls, 3m high side panels 4m in length

Sika joints between vertical panels

Dowels inserted at base of panels and grouted into position (see construction detail)

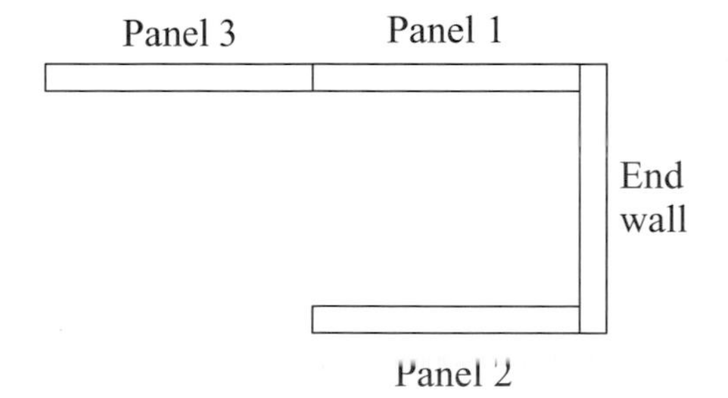

Precast walls in position

Position first wide-slab roof panel

Ditto–working across the roof.

Completed roof slab–100mm in-situ concrete to be placed over.

Day 2

Erect remaining wall panels to basement area. Commence fixing of floor units at one end of basement.

Day 3

Erect floor panels to basement roof area.

Complete basement walls and roof slab erected in 3-day period.

Day 4

Prepare formwork to edge of roof slab.

Day 5

Concrete roof slab

Part completed basement roof

Precast floor web sites

bison.co.uk	(Bison Concrete)
heidelbergcement.com	(Hanson Precast)
acpconcrete.co.uk	(Thomas Armstrong)

CHAPTER SIX

HANDLING CONCRETE

6.0 Overview

A range of concrete handling techniques are illustrated with explanatory notes on site situations. These include:

- placing concrete by direct discharge into a garage floor slab
- two situations of placing concrete using an excavator bucket
- a wide variety of skip types use with mobile and tower cranes.
  The use of skips in placing ready-mixed concrete is outlined for a variety of operations.
- the use of wet storage hoppers on site–this process is used mainly for construction operations such as piling. This is further illustrated in the foundations and piling section.

  The use of wet storage hoppers is also an economical consideration on concrete framed projects where the subcontractor places the concrete using his own static concrete pump in a wet hopper situation.
- a wide range of concrete pumping applications–these include the use of concrete mast pumps, static pumps and mobile pumps.

  A comparative cost study is shown, indicating an average cost of placing concrete to be approx. £21 pounds per cubic metre, altogether concrete pump hire charges vary widely, so it is difficult to accurately assess the real cost of pumping concrete.

Handling concrete–the easy way

6.1 Handling concrete

Various methods of handling and placing concrete are illustrated from practice.

Direct discharge

Direct discharge from ready-mixed concrete lorry alongside the foundations or floor slab. This is achieved by using the chutes attached to the rear of the delivery vehicle. An example of placing concrete in a small garage floor slab is shown.

Using an excavator bucket

The excavator is used to transport the concrete a short distance from the ready-mix delivery point.

An example of placing concrete in a tower crane foundation base is shown, and an excavator being used to place concrete in strip foundations.

Use of cranes and skips

A popular method of handling concrete in small or medium-sized pours is to use a crane and a variety of skip types.

On the continent, extensive use is made of site-mixed concrete using a site batching plant. In the UK we appear to have 'lost the art' of site mixing and prefer to use a ready-mixed supply.

Site batching plants are normally located within reach of a tower crane and the use of a wide variety of skip types are used.

The placing of concrete using a side discharge skip is illustrated using a mobile crane to place the concrete in a wall pour.

Use of wet hoppers

Wet concrete storage hoppers with a fixed concrete pump are extensively used for piling and large floor or wall pours. The wet storage hoppers are continuously fed with ready-mixed concrete. Concrete may then be pumped from the wet hoppers as and when required on site. Site-based wet hoppers are a more economical choice on a large concreting project, rather than hiring mobile concrete pumps.

Concrete pumping

A variety of concrete pumping situations are illustrated using mast-mounted pumps, through to static and mobile pump arrangements.

Pumping is the most-cost effective way to place large volumes of concrete. Illustrations are shown for placing concrete in foundation beams, large basement floors and floor slabs in mutli-storey buildings.

Concrete pumps cost between £350 and £600 per day to hire, for a small pour of 10 c.m. or so, this is an expensive solution, but it may be the only one.

6.2 Identification of hazards

Method statements and risk assessment when handling and placing concrete should address the following.

Likely hazards

- Moving plant–cranes and skips –Use site hazard board to highlight planned daily operations.
- Access to the placing point –Establish well-signed access routes for operatives. Erect barriers to guide operatives clear of work area.
- Vibration/vibrator problems
- Collapse of formwork –Formwork should be checked for stability before commencing a pour.
- Objects falling
- Working with concrete –Try to avoid dropping wet concrete from a height and creating splashing.
  –contact with eyes
  –cement burn to skin
  –mould oil

Control measures

- Full personel protection equipment
  including –eye protection
  –safety boots
  –gloves
- Use of banksman and radio contact with plant operatives
  –crane driver
  –ready-mixed concrete operators
  –pump man
- Providing adequate safe access to the placing point

Separate hazard assessments should be provided for concrete operations.

6.3 Concrete placing by direct discharge

Direct discharge into wheelbarrow

Archimedes screw mixing arrangement at rear of mixer (in extended conveyor arm).

Placing concrete directly from the ready-mixed concrete vehicle is the most economical method of placing concrete. This is advantageous when the delivery vehicle may be positioned alongside the foundation or slab pour.

To assist placing, additives may be added to the concrete mix to aid flow properties. Self levelling additives for floor slabs my also be specified.

6.4 Concrete placing by excavator

Concreting of foundation base for a tower crane

The task involves the constructing of a large pad foundation for the base of a tower crane adjacent to a large basement. Images show of ready-mixed concrete being placed in the excavator bucket and being discharged into the base.

Alternative concrete handling methods could have been considered such as direct discharge via the extended chutes provided on the vehicle, but this is dependent on access conditions.

Concrete pumping may be an economical proposal when a number of smaller pours on the same day are organised, e.g. a base pour, wall pour and column pour etc.

Placing concrete with an excavator bucket

This is not the normal or even an economical way of placing concrete, but when it is the only plant you have on site, then it is a solution.

On the smaller site the use of an excavator may be considered for transporting the concrete across the site and placing it in the trench. It is a common practice on the smaller projects where local subcontractors are being employed.

6.5 Concrete placing by crane and skips

Types of skip

A wide variety of concrete skip types are available for use in concreting operations. These include bottom-discharge skips, roll-over skips and skips fitted with pouring pipes (usually termed tremi pipes or an elephant's trunk). Tremi pipes are used to limit the 'drop' of concrete, as it is often considered that concrete segregates if dropped from a height of more than 1.2 metres. Using skips for placing concrete is a slow process as it may take up to one hour to place 6 c.m. of concrete, but it may be the only solution.

Concrete placing using a side-discharge skip

Concrete skips are widely used for the placing of concrete. Bottom and side-discharge skips are available and are used with mobile cranes. They are available with flexible discharge trunks. Placing is relatively slow and expensive in both labour and plant costs.

Skip arrangements are normally used for relatively small and medium-sized pours, e.g. in the placing of concrete in columns and suspended beams. Alternatively, a number of small pours could be organised to be completed in a single visit in order to make pumping a more economical proposition. A minimum of three labourers are required to undertake the placing operation: one placing the concrete, one vibrating the concrete and one acting as a banksman for the team. The concreting of a full-height wall pour is illustrated.

Skip types used with site-mixed concrete

A further range of skip types are illustrated. On the continent (especially in France, Germany and Italy) site-mixed concrete is extensively used. The continentals do not appear to have 'lost the art' of producing good-quality concrete on site. In these circumstances, use is made of a wide range of concrete skip types. Extensive site precasting is also undertaken on European projects.

Tower crane and skip concreting the 12th-floor lift of a sliding formwork pour

Concrete skip fitted with extended trunk placing concrete in 300mm thick wall.

6.6 Concrete placing via wet hoppers

Placing concrete via a wet hopper and concrete pump unit

The use of wet storage hoppers is common practice when supplying concrete to operations such as bored or driven piling. The ready-mixed concrete can be stored on site in the wet hopper until it is required for placing in the pile shaft. This speeds up the construction sequence because when a pile shaft is ready for concreting, the concrete is readily available.

Wet storage hoppers are simply the mixing drum from a ready-mixed concrete vehicle mounted on a platform that can be positioned on site. A concrete pumping unit is located in front of the mixer to facilitate the pumping operation.

6.7 Concrete pumping

Use of a mast pump and mobile pump

Concrete pumping situations

Concrete pumping is the most efficient way of placing large quantities of ready-mixed concrete. The size and the location of the pour are critical factors. Where a number of smaller pours in different site locations may be organised in the same day, it may also prove to be an economical option.

Concrete pumping requires careful planning. Time is required to set the pump up in the correct location and to lay out hoses/pipelines and equipment. Pumping operations normally require more labour as the concrete requires levelling and finishing. Pumping rates of 20–30 c.m.per hour may be achieved, depending on the access to the pour. The pumps are normally hired out on a half or full day basis, with rates at £350–600 per day. For small pours, pumping is not a cheap option, as previously explained.

Pump attached to readymixed delivery vehicle in Denmark

A Concrete pumping unit fitted to a readymixed concrete delivery vehicle. The pump discharge hopper is located at the rear of the machine. the arrangement is ideal for pumping relatively small concrete pours

The pump boom unit is located between the rear drivers cab and the mixer drum

Concrete pumping to foundation beams

Completed group of piles

Plywood formwork to ground beams

Concrete pump in position

St Helens project

Foundations to five-storey block of residential flats

Concrete ground beams on precast concrete piled foundation. Ground beam section 600mm wide by 900mm deep. Plywood formwork to beam sides shown

A well-organised and tidy project

First-class relationships developed between company and subcontractors

Concrete pumping to bridge deck

Pumping boom in position

Concrete pump unit

Placing and leveling concrete

Paddle float for finishing slab

A 30 c.m. pour to a bridge deck was completed in 4 hours

- Four labourers placing and finishing concrete plus a supervisor
  Output rate from pump approx. 8 c.m. per hour
- On placing concrete in strip foundations and large slabs, extensive use is made of laser levelling equipment. Control of surface levels can be easily maintained within + or –3mm

Concrete pumping to a muti-storey block

Placing concrete by mobile concrete pump to the fourth floor of a building–80 c.m. floor pour in progress

Concrete pump in position for concrete slab pour on seven -storey building

Cost of concrete pumping

Situation one–concreting a strip foundation

		£
Plant costs		
Hire of pump	350	
Travelling time	100	
Labour costs		450
3 labourers at £100/day		300
Total labour & plant costs		750
Concrete placed 30 c.m.		
Labour & plant cost per c.m.		£25.00

Situation two–concreting a bridge deck

		£
Plant costs		
Hire of pump	600	
Misc. plant	50	
Labour costs		650
4 labourers at £100/day		400
Total labour & plant costs		1050
Concrete placed 50 c.m.		
Labour & plant cost per c.m.		£21.00

Situation three–concreting a tank base

		£
Plant costs		
Hire of pump	600	
Misc. plant	50	
Labour costs		650
4 labourers at £100/day		400
Total labour & plant costs		1050
Concrete placed 60 c.m.		
Labour & plant cost per c.m.		£17.50

Average costs (three projects) £21.00

6.8 Basement slab case study

Description of project

The basement is 50m × 38m long and forms the car park area for two blocks of apartments. The plan illustrates the layout of the basement and shows the access road for construction traffic.

Sequence of pouring the basement slab

The project specification requires the contractor to submit proposals for concreting the basement floor and associated foundation bases to the seven-storey apartment blocks. The slab thickness is specified as 450mm.

Proposals included by the contractor contained a layout plan showing the proposed floor pouring sequence. The joint types to be used in the floor construction are indicated as required by the consulting engineers. Method statements were also submitted by the contactor together with an assessment of construction hazards (risks were identified within the written method statements).

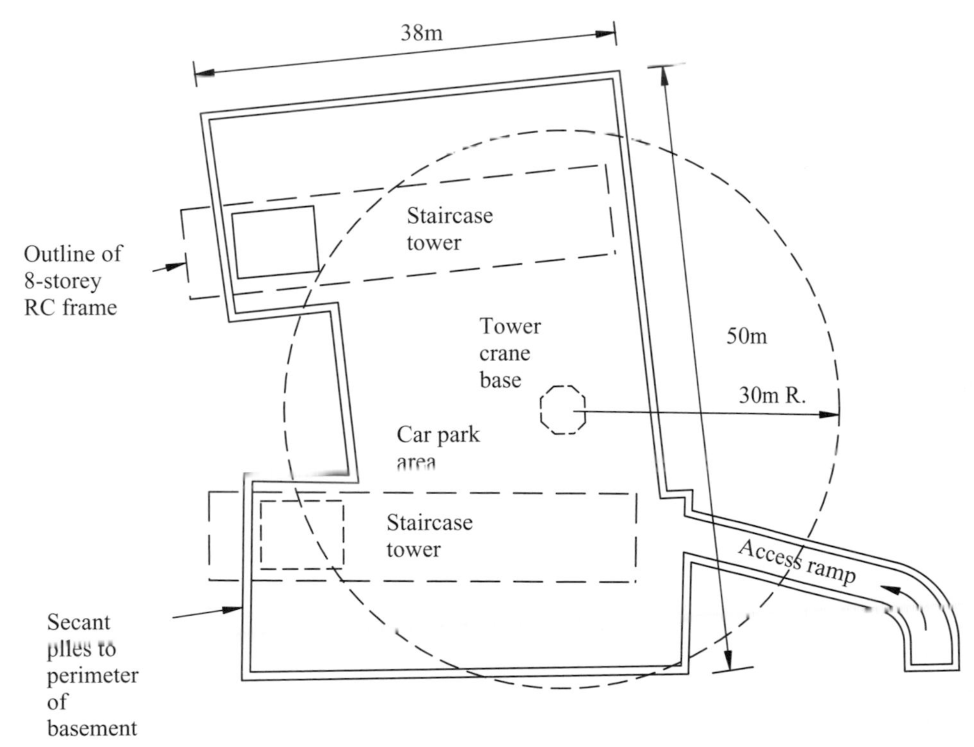

BASEMENT PLAN

Basement floor pour–construction sequence

View of basement floor showing 5m wide floor bays with pouring sequence indicated
Concreting using pumped concrete placing from static pump, based on using wet hoppers

Longitudinal stop-end

Reinforcement bars tied into bored pile wall. Bentonite sheet treatment to face of bored pile wall

Concreting of floors undertaken by static pump working in conjunction with wet hoppers

5m wide bays

View of completed pour

Overall basement size 50m x 38m

Basement slab 450mm thick

Concrete placing undertaken using six labourers with steel fixer and carpenter on standby

Concrete quantities in a single bay pour = 25 x 5 x 0.45 = 57 c.m. per bay

Mechanical screed board used for obtaining smooth finish to floor slab surface

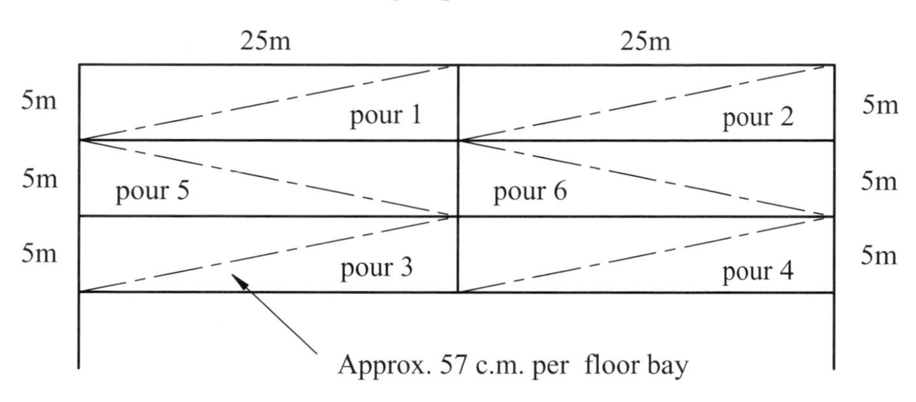

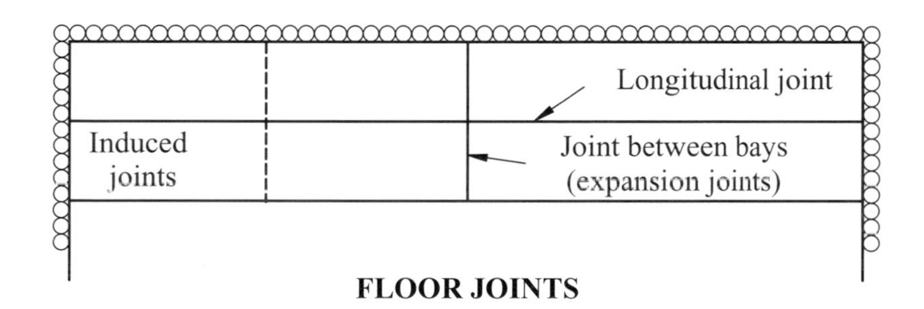

FLOOR JOINTS

Recording pouring sequences on the project

It is essential for site records to record the dates the floor pours arc undcrtaken. This responsibility is usually carried out by the assistant site manager or a site engineer. Floor pans require to be colour coded to indicate the sequence of pouring and contain reference to the concrete test cubes taken from each bay.

This may also be extended to all concrete pours i.e. floor slabs, columns, beams and wall pours, which can be readily marked on floor plans and elevations.

6.9 Large floor slab pour

Description of project

The task involves the placing of a concrete slab 250mm in thickness. The building floor plan indicates the layout of a three-bay pour.

Pour 1 - 28m x 30m x 250mm = 210 c.m. 1 day pour

Pour 2 - 28m x 30m x 250mm = 210 c.m. 1 day pour

Pour 3 - 28m x 20m x 250mm = 140 c.m. 1 day pour

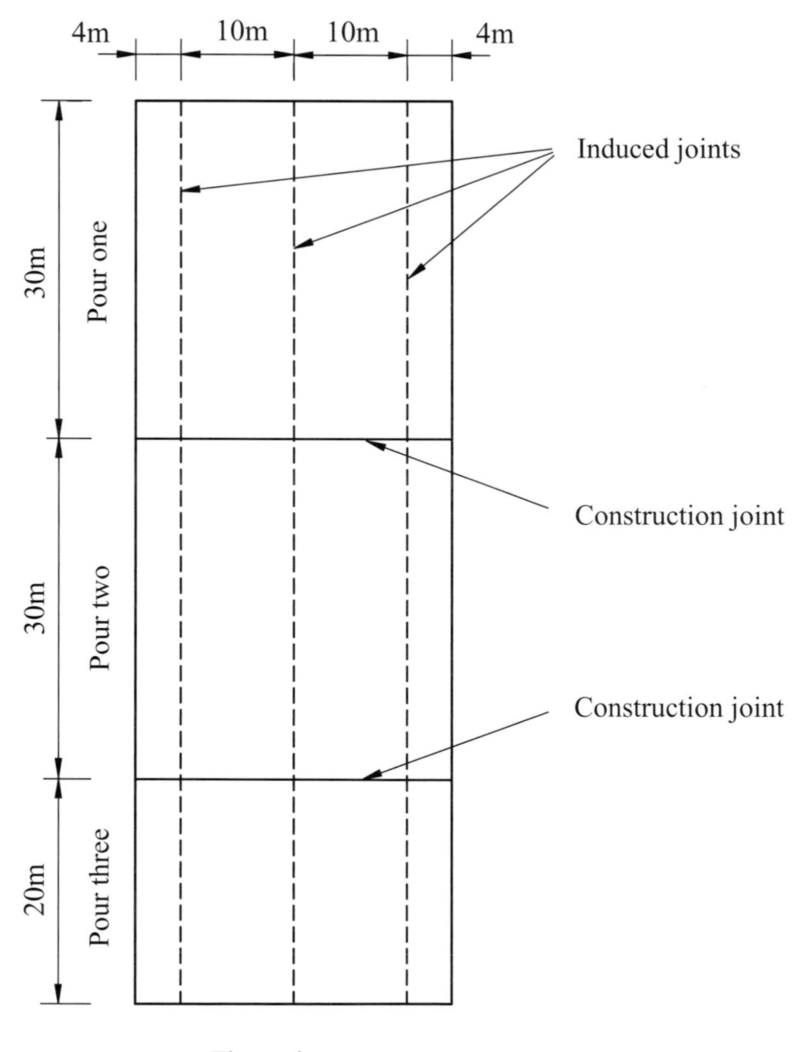

Floor plan

Floor joint details

Large floor pours are a common practice in today's construction industry. The contractor requires approval of the proposed pouring sequence from the client's engineering consultant. This relates mainly to the size of pour and the joint layout to be incorporated in the floor slab.

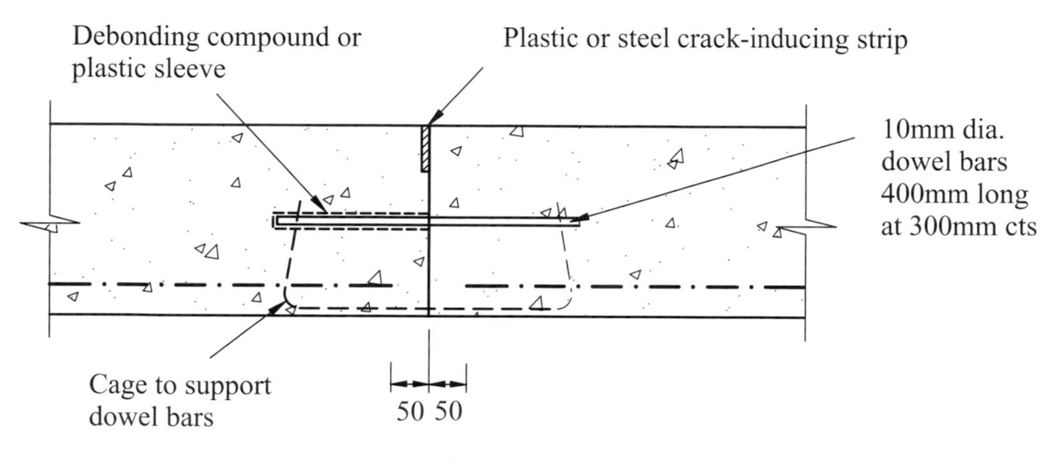

Construction joint

Details of the construction joint (between adjacent pours) and the induced joints to be cut in the slab surface within 24 hours of the floor pour

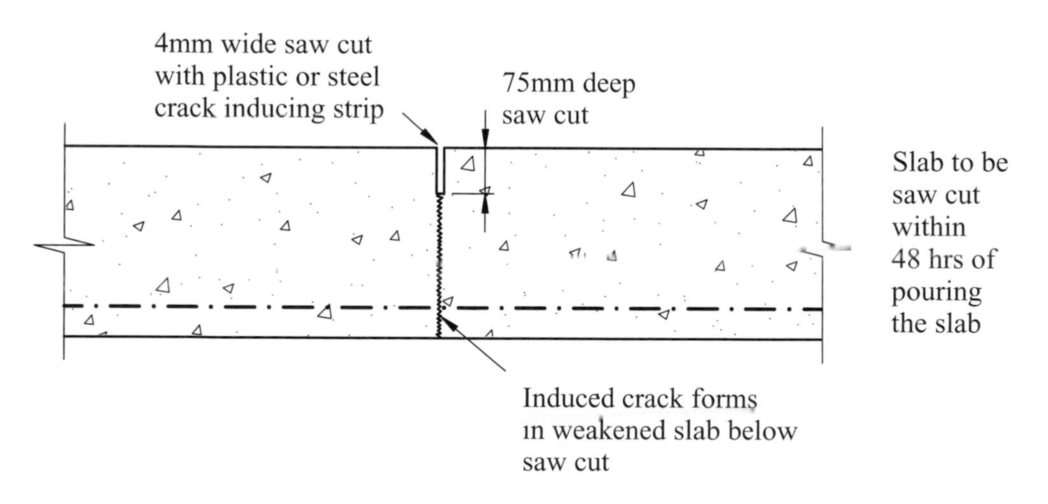

Induced joint (max 10m centres)

CHAPTER SEVEN

BUILDING FRAMES

7.0 Overview

Building frames in concrete construction have been classified as column and beam frames (traditional reinforced concrete frames), flat slab construction and crosswall construction. These incorporate both in-situ and precast concrete components.

Traditional concrete frames incorporating downstand beams and in-situ floor construction appear to have lost favour as a modern form of construction. No projects incorporating traditional construction could be observed in the north of England.

Flat slab construction is used on a number of projects in both Manchester and Liverpool. This form of construction incorporates the use of tableform formwork systems which are illustrated in the construction of two seven-storey residential blocks.

Crosswall construction is widely used in both the UK and mainland Europe. This method of construction may be in-situ walls and floors or a combination of precast wide slab floors or precast plank systems. Images are shown of projects in the UK, the Netherlands and Spain.
Detailed construction sequences are shown of a ten-storey precast crosswall frame in Manchester.

Formwork systems for flat slabs and tableforms are also illustrated in Chapter 8.

Crosswall construction to the twelve storey Hilton Hotel in the Liverpool One complex.

7.1 Building frame types

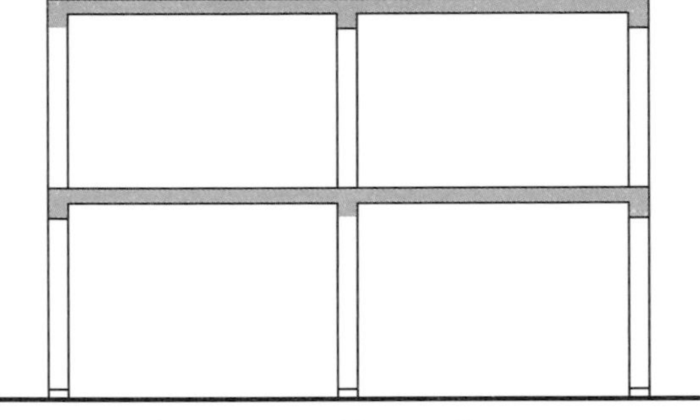

Column and beam frame
beam and floor slab

Square grid of floor beams or a rectangular grid arrangement. Downstand beams formed as integral part of slab or cast separately to the floor.
Maximum economical span 5–6 metres.
Wide variety of cladding arrangements available.

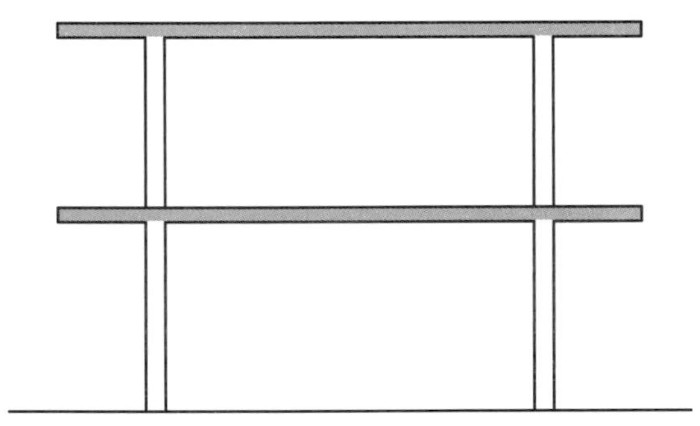

Flat slab frame

In this form of construction the slab is of uniform thickness.
Column spacing 5–8 metres.
A drop floor may be incorporated, or the column head may be thickened.
Columns may be set on face of building or set back.

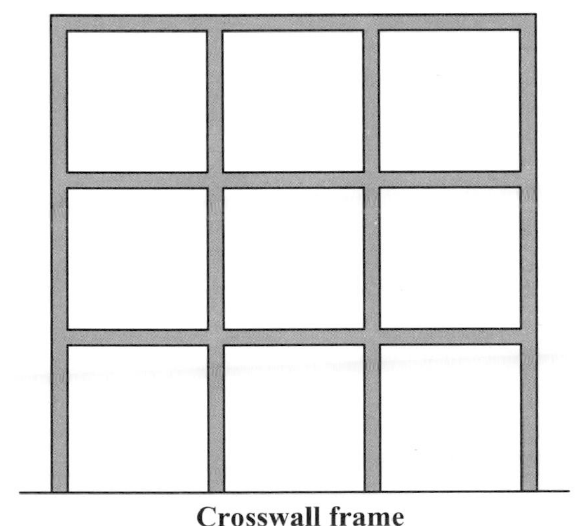

Crosswall frame

Consists of in-situ concrete walls and floors. Span between crosswalls 5–7 metres.
Suitable for buildings with identical layout plans on each floor.
The wall and floor units may be of precast construction incorporating wide slab units.
Plank floors may also be introduced to reduce formwork costs.

Flat slab construction

Seven-storey in-situ concrete frame showing table forms at the sixth floor of the building

Crosswall frame

In-situ concrete crosswall construction on the Hilton Hotel in the Liverpool One development.

7.2 In-situ concrete frame with downstand beams

In-situ concrete columns and downstand beams supporting a 200mm thick in-situ floor slab. Two storey apartment block in Europe

A variety of in-situ column and beam sizes shown varying from 400 x 250mm to 600 x 250mm in section

Traditional concrete frame

In-situ concrete frame with downstand beams

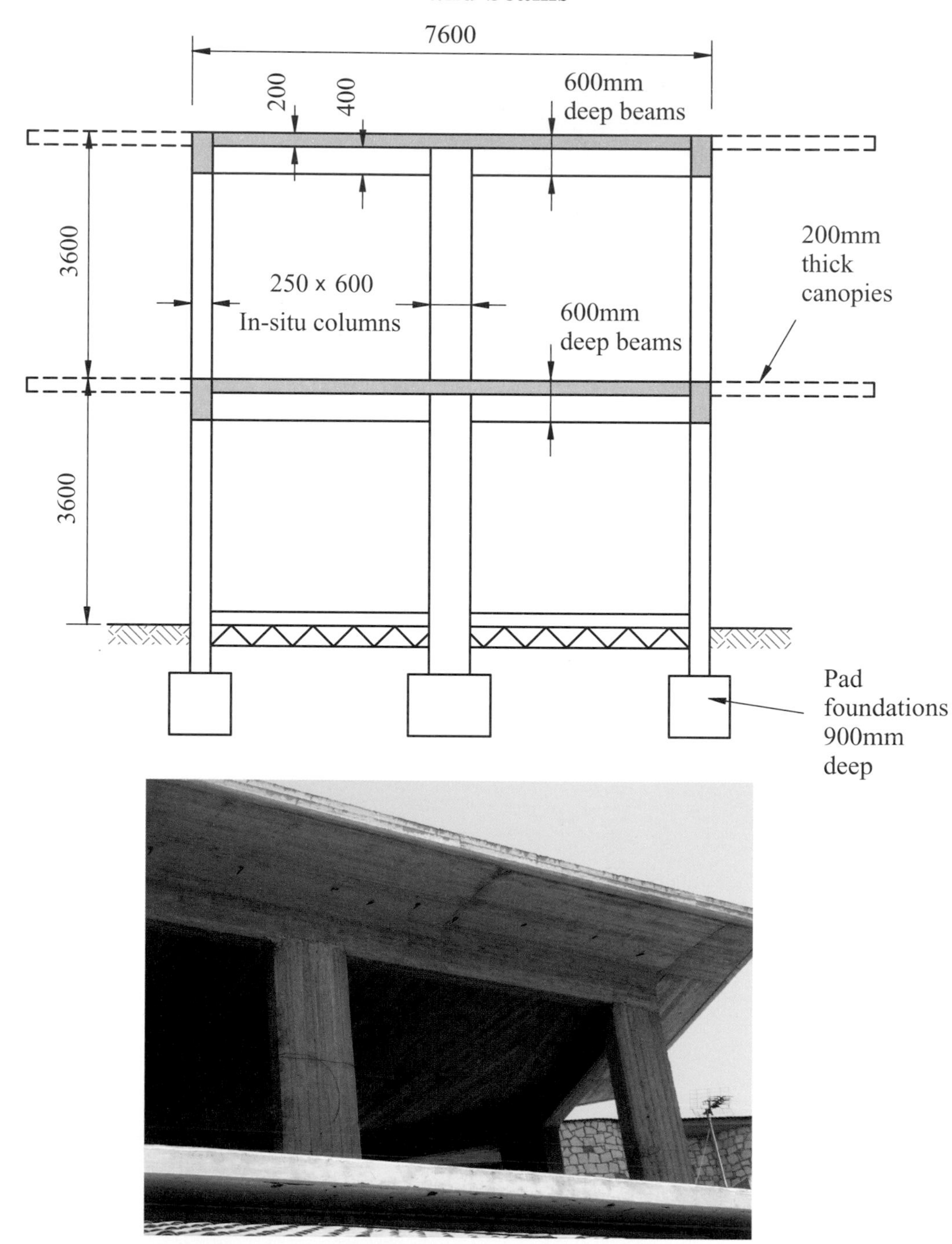

Internal image of column and beam arrangement

7.3 Flat slab construction

Simply consists of a series of in-situ concrete columns supporting a flat concrete slab. This form of frame construction lends itself to the use of tableform platforms for forming the support to the in-situ concrete floor slabs.

The columns may be set back from the face of the frame. Curtain walling systems may be used to provide an ultra-modern glass block appearance to the finished building.

The lower picture shows back-proping to the lower floors using extendable aluminium props prior to fixing external claddings.

Tableform construction incorporating post-tensioned in-situ concrete floors

Post-tensioned cables in position
in in-situ slab

Curved storey-height panels to
flat slab building

Flat slab construction

Completed flat slab floor area

Flat slab construction

Flat slab design is ideally suited to residential developments involving multi-storey construction. The use of tableform formwork systems is common practice; where the tableforms are lifted from one floor to another during the construction cycle.

The project illustrated is a seven-storey block of flats with a seven-storey staircase core. The in-situ concrete core was constructed using sliding formwork. Details of the sliding formwork as illustrated in chapter 9.

The building columns may be located on the building's external face or set back to accommodate the cladding system.

Tableforms being handled by tower crane into position onto the next floor

Section of floor being prepared for concreting
Floor pours all undertaken by mobile concrete pumps

**Two seven-storey residential blocks
Castlefield project**

Prestige residential development in the Castlefield Basin area of Manchester. Development consists of two seven-storey blocks with basement car parking. The project is sited within the Castlefield canal basin. The reinforced concrete frames are of flat slab construction with aluminium curtain walling to external elevations.

The site is surrounded by canals which were constructed in the early 1800s.

The new residential developments improve the inner city areas and regenerate interest in our canal heritage.

7.4 Crosswall construction

Form of construction used for main frame of multi-storey hotels/high-rise flats and medium-rise residential blocks.

Consists of in-situ concrete or precast crosswalls at 6–8m centres forming structural walls between adjacent units.

Floor construction may be in-situ or precast construction. The external cladding may be in-fill panels fixed to the ends of the crosswalls or a curtain walling system. Crosswall construction is extensively used for low-rise housing construction in Europe.

Crosswall construction–sequence of work

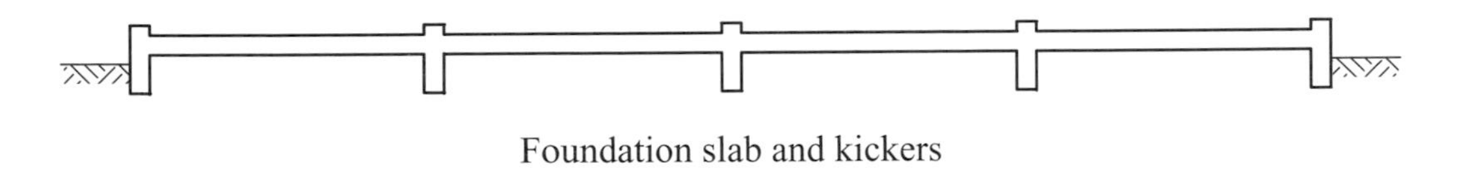

Foundation slab and kickers

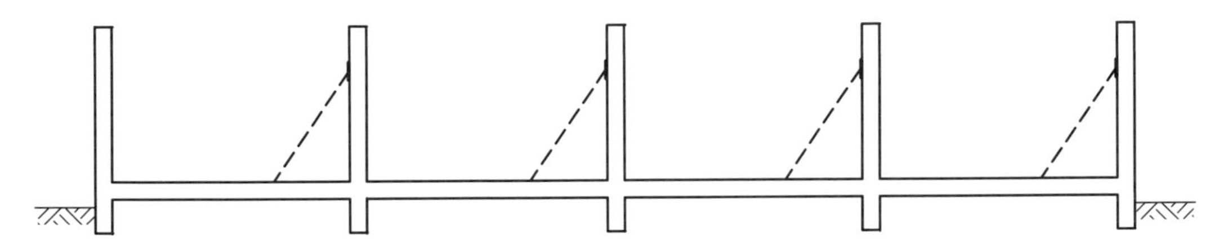

Erect in-situ concrete walls to first floor level

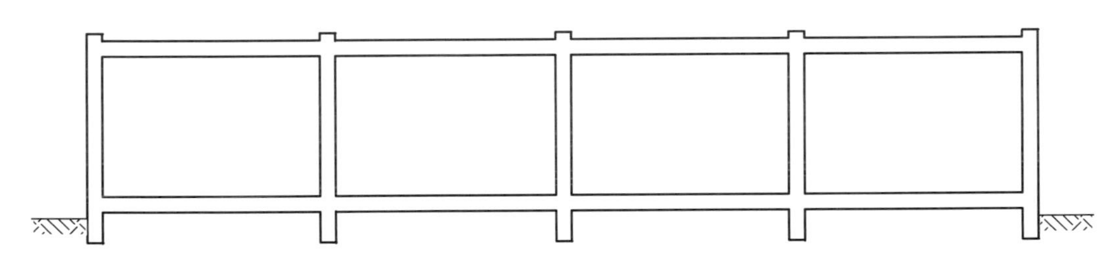

Construct in-situ/precast first floor slab

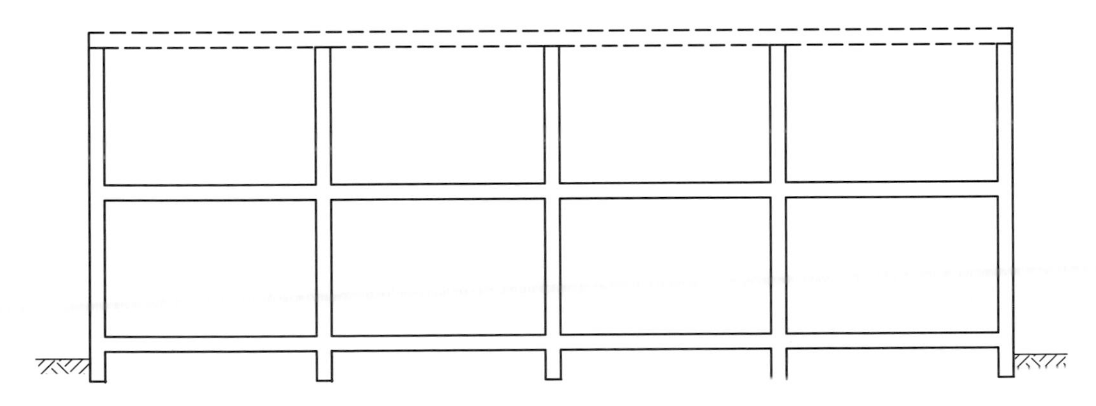

Erect walls to second floor level

Crosswall construction–connection details

Connection between precast concrete crosswalls and precast floors.

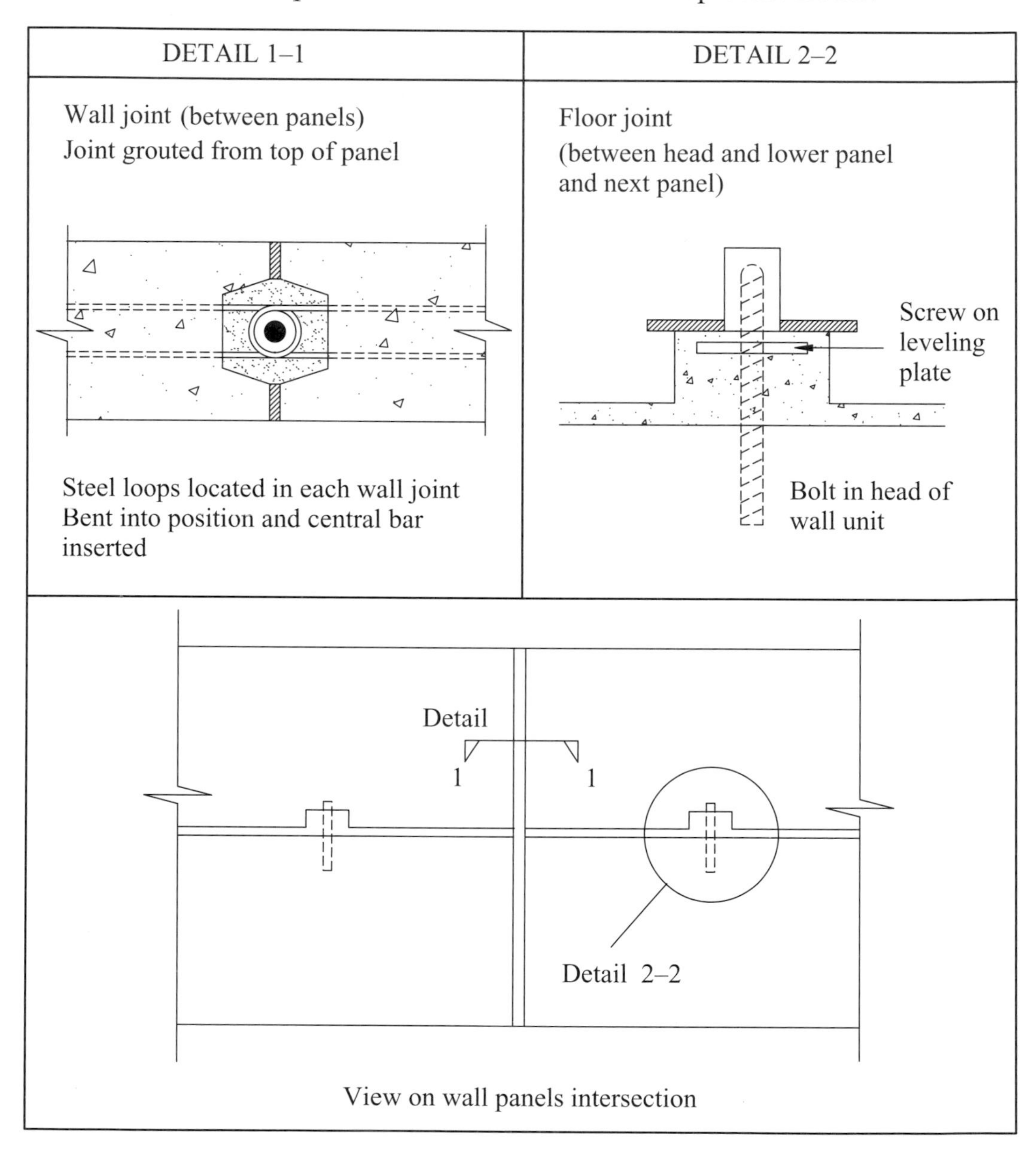

Crosswall construction in europe

Crosswall construction in mainland Europe is used for medium and high-rise flats. The use of in-situ floors and crosswalls is common practice. For medium-rise blocks, the use of in-situ crosswalls and precast plank floors is often specified. The plank floors act as permanent floor soffit formwork and are supported on temporary props as illustrated above.

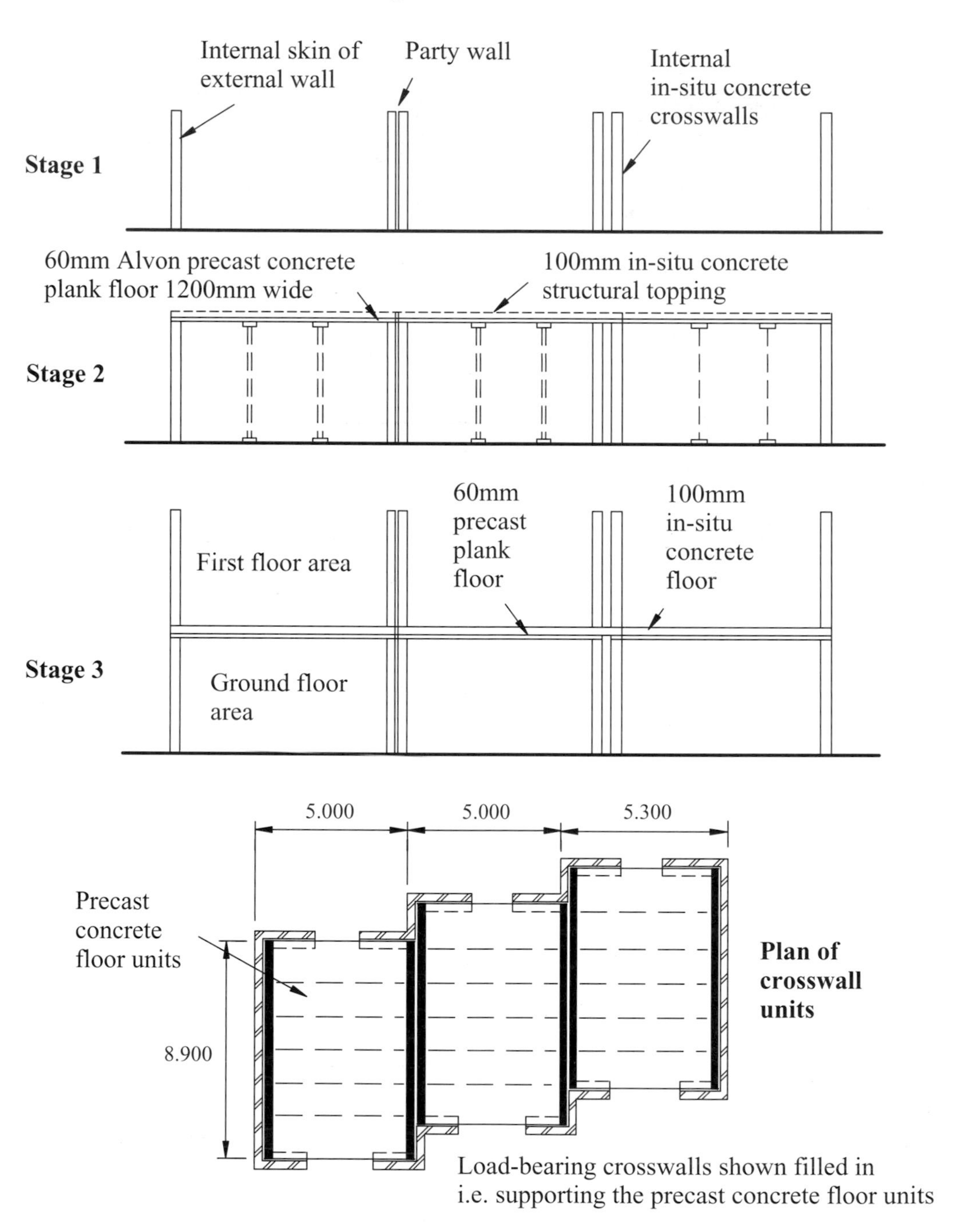
Sequence of work
Internal skin of external wall
Party wall
Internal in-situ concrete crosswalls
Stage 1
60mm Alvon precast concrete plank floor 1200mm wide
100mm in-situ concrete structural topping
Stage 2
First floor area
60mm precast plank floor
100mm in-situ concrete floor
Stage 3
Ground floor area
5.000
5.000
5.300
Precast concrete floor units
8.900
Plan of crosswall units
Load-bearing crosswalls shown filled in i.e. supporting the precast concrete floor units

Crosswall construction in Spain

A large number of medium and high-rise hotel blocks are of in-situ concrete crosswall construction.

The quality of the concrete structural frames varies widely. Construction defects observed include:

- lack of cover on reinforcement
- extensive honeycombing on the face of walls and beams
- spalling of concrete at arrises of beams
- cracking of balcony walls
- failure of concrete screeds and extensive cracking of floors and finishes
- formwork support systems inadequate, often unsafe
- lack of quality control and inspection.
- lack of safety procedures in general

A block of 20 apartments under construction. The quality of the concrete construction leaves a 'lot to be desired'.

A completed apartment block of twenty units. They have certainly employed good concrete patchers, renderers and painters here.

7.5 Crossawall construction–Eastlands project

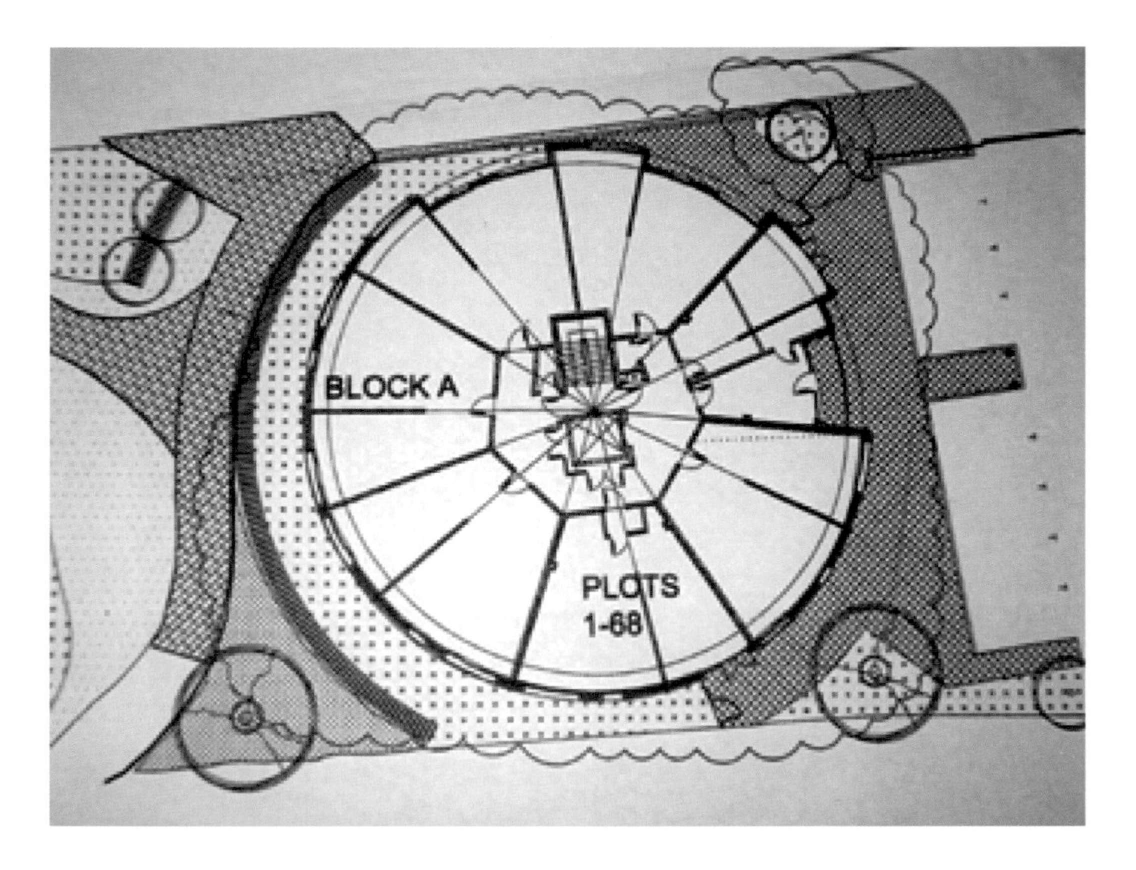

Description of project

The project involves the construction of two circular ten storey precast buildings. The value of the project is about £20m. The circular blocks contain seven flats per floor using crosswall construction.

A central building core contains lifts and staircase access facilities to each floor. The construction cycle time per floor is 6 days. This allows for a 10-week programme to construct the building frame on each block. The frame supply and erection work is awarded as a work package.

A management case study on the project is presented in *Constructing, Planning, Programming and Control,* Cooke & Williams, 3rd Edition.

The erection sequence per floor, in days:

Erect perimeter wall	1
Erect crosswalls	1
Erect protection	1
Erect floor nets	1
Erect floor units	1
Grout up	1
Cycle/floor	6 days

A programme for the erection sequence for two floors of the building is shown on the precast erection programme diagram

The provision of edge protection and of safety nets prior to the floor erection is an essential safety requirement.

Crosswall construction

The edge protection to the wall perimeter is shown together with floor nets in position.

The floor erection needed careful co-ordination with the precast factory due to all floor units in one bay being tapered to accommodate the floor plan shape.

After fixing the floor units the wall and floor joints were grouted prior to erecting the walls to the next floor.

Eastlands project programme

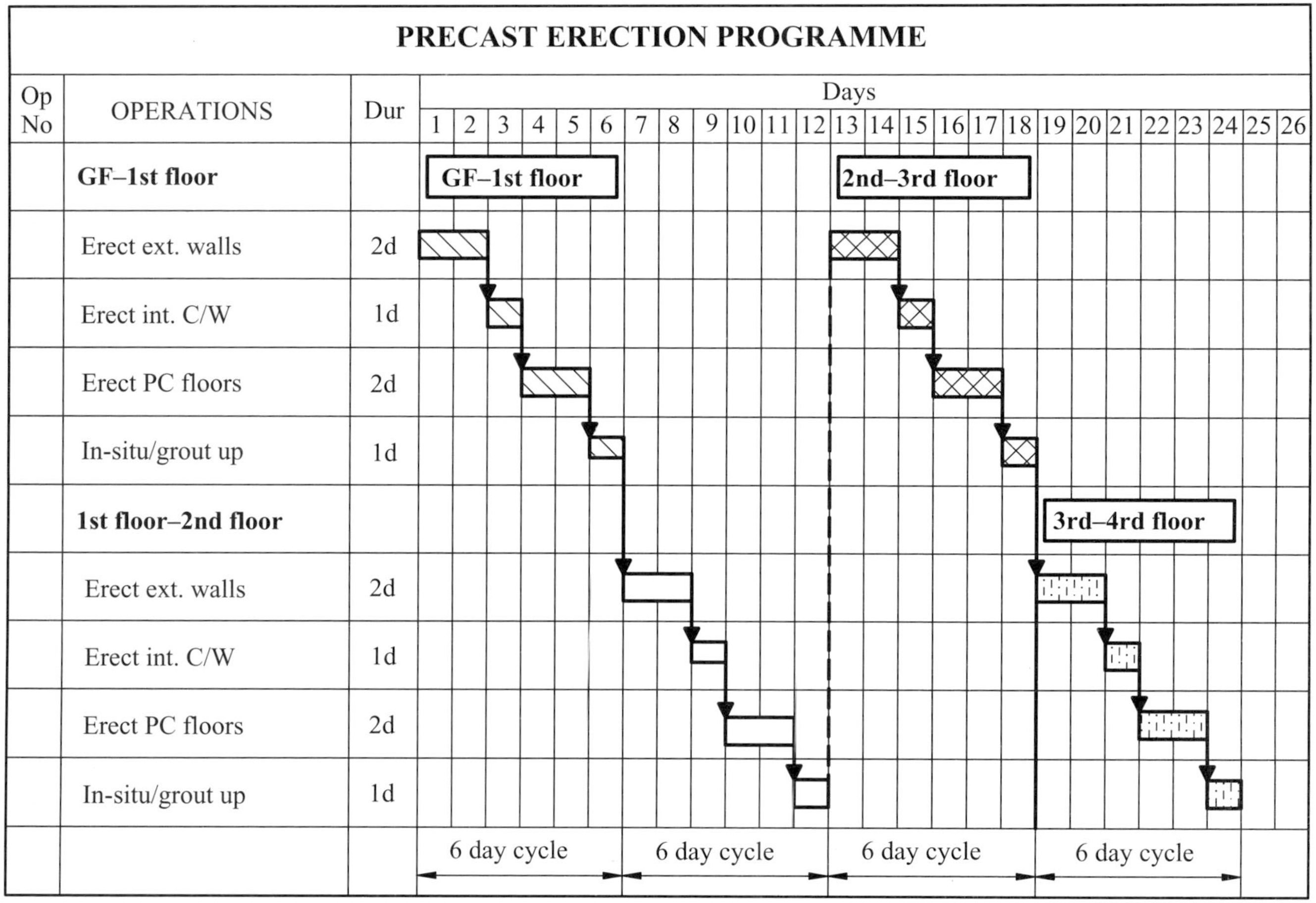
PRECAST ERECTION PROGRAMME
Op No
OPERATIONS
Dur
Days
1 2 3 4 5 6 7 8 9 10 11 12 13 14 15 16 17 18 19 20 21 22 23 24 25 26
GF–1st floor
Erect ext. walls 2d
Erect int. C/W 1d
Erect PC floors 2d
In-situ/grout up 1d
1st floor–2nd floor
Erect ext. walls 2d
Erect int. C/W 1d
Erect PC floors 2d
In-situ/grout up 1d
GF–1st floor
2nd–3rd floor
3rd–4rd floor
6 day cycle
6 day cycle
6 day cycle
6 day cycle

An extract from the precast programme is illustrated for the floors of the building.

The overall contract period was based on a 52-week contract which was successfully achieved.

A project of this type requires a good relationship to be developed between the main contractor and the work package subcontractors.

CHAPTER EIGHT

STEEL-FRAMED STRUCTURES AND ROOFS

Construction Practice. Brian Cooke

8.0 Overview

Steel-framed structures and roofs

18.1 Introduction to skeleton frames

18.2 Steel portal frame

18.3 Cellular beam frames

18.4 Space frames

18.5 Arched tubular roof

18.6 Cable-stayed buildings

18.7 Cable-stayed roof to Metro station

18.8 Bow-string roof to entrance foyer

18.8 Glulam portal frame–church project

18.9 Glulam portal frame–Sheffield winter gardens project

This section is intended to show images of a range of steel-framed structures which are accessible to students, lecturers and the public alike.

These include a space frame structure at Manchester Airport Terminal 2, the arched roof forms of the atrium area of the Richmond Building at Bradford University, the cable-stayed car sales building at British Car Auctions on Hyde Road, Manchester, and the cable-stayed Metro station on Oldham Road, Manchester for the new Metro link to Oldham.

All these sites can be accessed by the public. Students should be encouraged to integrate images of construction projects into construction assignments whenever possible.

View of rectangular skeleton frame

8.1 Introduction to skeleton frames

This consists of a steel-framed building with vertical steel columns and horizontal beams. The framework may be constructed on a rectangular grid layout to support the floors, roof and walls of the building which are all attached to the frame. Columns may be formed of I beams or tubular steel.

Floor construction usually incorporates metal decking which is shot fired to steel beams. This assists in stabilising the frame.

Cantilever frames may also be incorporated in the design (see the Mann Island frame images on page 207).

The steel frame may be further stiffened by tying the steelwork to in-situ concrete vertical lift shafts, as illustrated.

This image shows a rectangular skeleton frame constructed of tubular columns and incorporating cellular beams.

Skeleton cantilevered frame

Building at 8th floor steelwork erected at one floor per 10 days

Mann Island office block Liverpool

Fifteen-storey skeleton frame with cantilevered floors. 500mm dia. steel columns (fabricated in Japan).

Metal deck floors positioned and concreted with a static mast pump located at the rear of the building.

Case study included in formwork chapter on the erection of sliding formwork to the staircase cores.

Cantilevered steelwork at each end of main building

Steel cap to columns at 3rd floor level

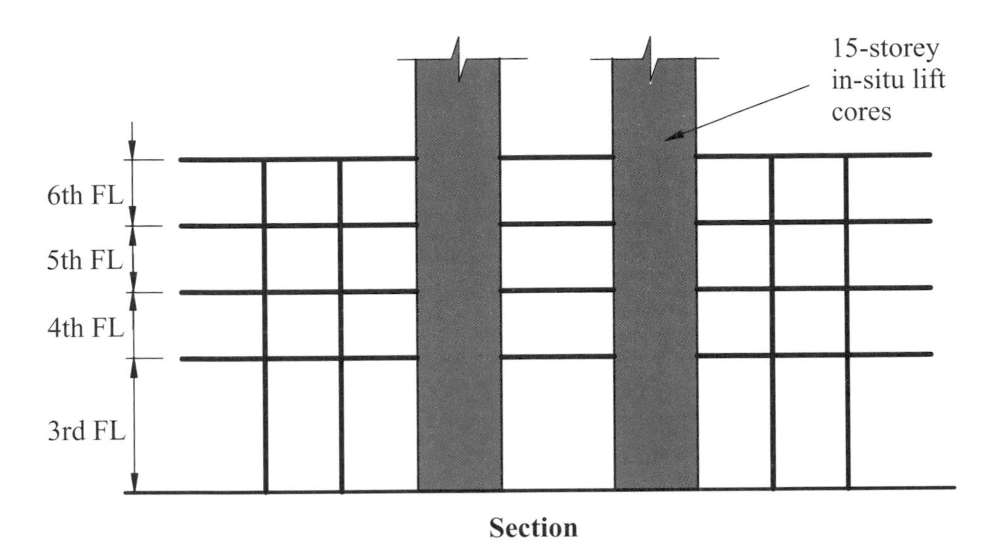

Section

Similar six-storey steel-framed building tied into in-situ concrete lift shafts. Note cantilevered floor over low-level building

SKELETON FRAMES

8.2 Steel portal frames

Steel portal frames are capable of spanning up to 60 metres ('short spans' up to 25m, 'medium spans' 16 - 35m and 'long spans' up to 60m). They are used for the construction of factories, warehouse buildings and leisure centres. Wall and roof bracing is normally required in selected bays and at the end of the building. Additional vertical columns may be introduced at the gables (wind posts) to support cladding on the end walls. Roof beams and columns are usually fabricated from rolled-steel sections, and Z purlins support the roof coverings. Claddings may consist of built-up cladding systems and fully insulated composite roof
systems.

Corus provide an excellent web site with teaching resources at Corusconstruction.com

Lattice portal frames

Lattice portal frames may be built up of an open grid of steel angles or tubular members. Pinned joints may be incorporated at the apex or base of the stanchion. One-pin, two-pin or three-pin arrangements are available. The pin joint at the stanchion reduces the bending moment at the support often resulting in a more economical foundation solutions. The stability of the lattice frame relies upon providing adequate side bracing and wind bracing. Portals are normally spaced at 6m centres.

A wide variety of wall and roof claddings are available, incorporating fully insulated sandwich panels. Single-ply roof systems are available and excellent web sites are provided by Kingspan, Ward Insulated Panels and Sarnafil products.

20m span steel roof

Traditional portal frames

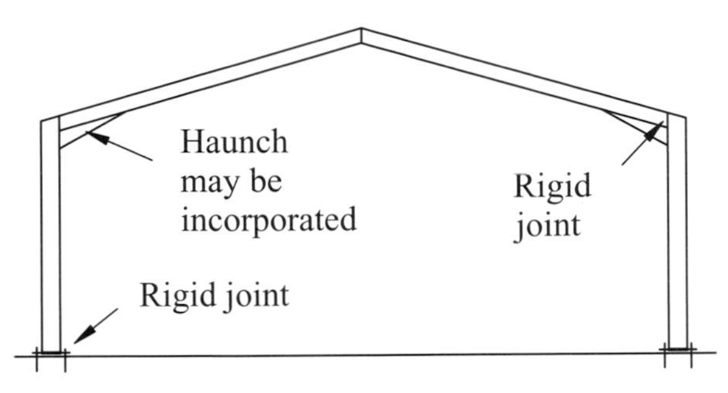

Riged portal frame

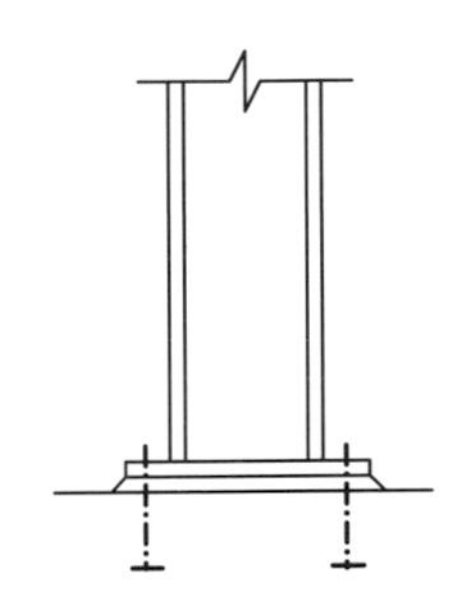

Rigid connection to base

Lattice portal frames

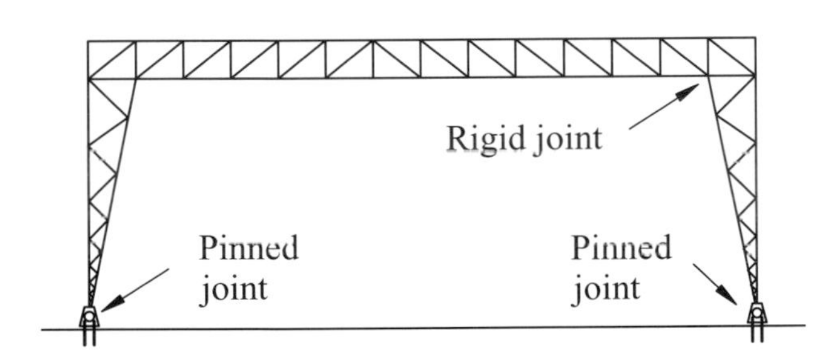

2-pin lattice frame

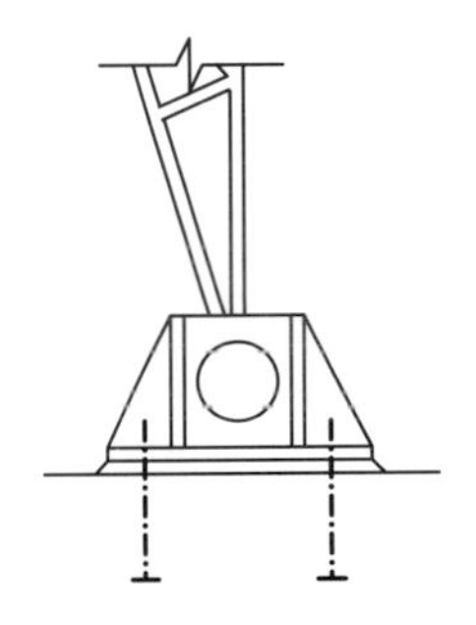

Pin joint at support

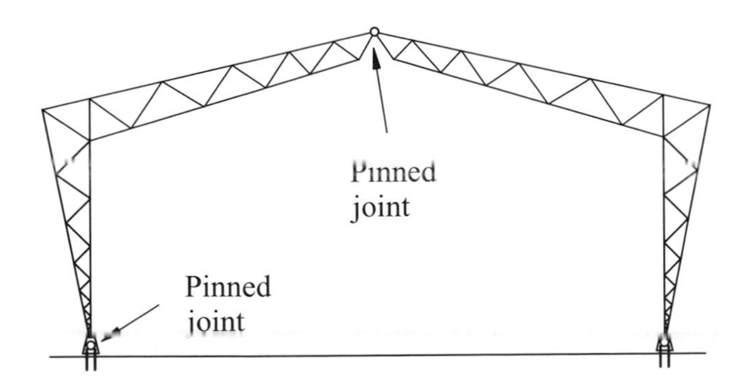

3-pin lattice frame

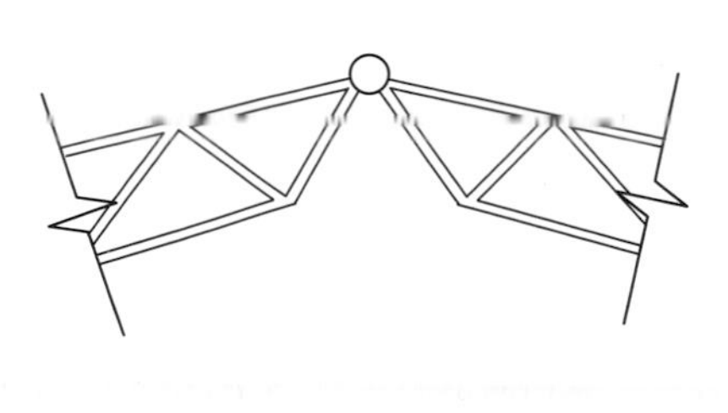

Pin joint at apex

Steel portal frame

Internal view of 15m span portal frame–Kingspan cladding to external walls and roof

Haunch joint between vertical column and roof beam

Apex joint at centre span–bolted connections and ties to Z purlins

Steel portal frame

Z purlins to roof at 1.5m centres–roof bracing to provide stability to gable frame

External wall–2m high brickwork with Kingspan wall cladding

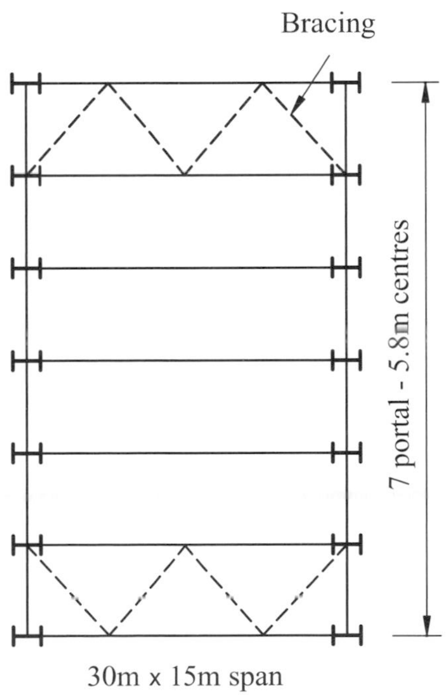

30m x 15m span
6 bay portal

8.3 Cellular beam frame

Description of frame

The tubular-framed building is 28m span and 80m in length. Bays are spaced at 5m centres with seven tubular columns supporting the main beam along the centre of the building. All columns are 325mm diameter supporting a series of 650mm deep roof beams. The central cellular beam is 900mm deep.

The contract value is approx. £700,000 and the erection period is 11 weeks.

View of eight completed bays of a sixteen bay frame

Rigid frame situation

The cellular beam frame has fixed base connections to the concrete foundation. The column base plate is fixed to the foundation base with four holding-down bolts as illustrated on the foundation images.

Commencement of frame erection

Part completed frame

Cellular beam frame

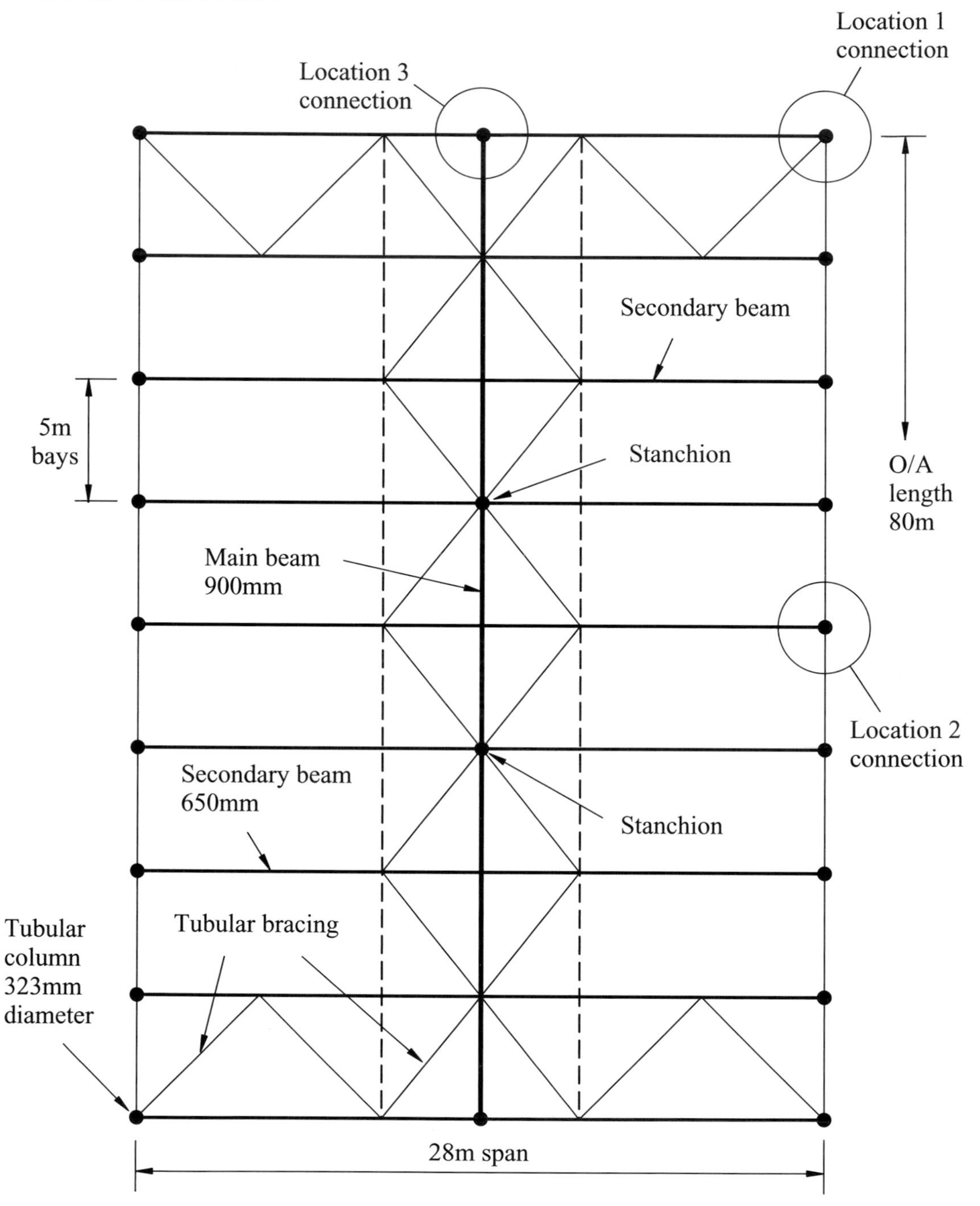

Plan of building (half plan)

Connection details

Location ① External corner connection

Frame connection at each external corner of the building shown.
The 325mm diameter circular column is connected to 200mm diameter tubular beams connecting the column heads.
The plate connections to intermediate columns is shown at location 2.

Location ② Secondary 650mm-deep beam connection.

Close-up of plate connection 2.

Location ③ Main beam connection

The frame arrangement is not a true portal frame as the main loading is transferred to a 900mm-deep cellular beam supported on columns located down the centre of the building.
The exposed cellular beams are used as a feature inside the completed building.

Cellular beam frame

Foundation detail

View of strip foundations showing holding-down bolt positions for main steel frame.

Position of steel levelling packing shown under steel column.

Steel circular column in position bolted to base. Steel levelling packings can be seen.

At each stansion position the centre line of the column base plate has been marked on the concrete base in two directions, i.e. in the blue marked areas.

Roof decking

The main cellular beams are positioned at 5m centres and the 10m-long insulated panels span every two bays.The surface of the panels may be finished in a site-applied single-ply plastic membrane finish.

View of insulated panels in position. The panel's screw fixings are fixed securely into top flange of the supporting steel beam.

Cellular beam frame

Roof decking

The insulated roof panels are Kingspan KS 1000 Top Span. The internal steel liner sheet is finished in a white polyester coating. This provides an attractive ceiling finish to the workshop area below.

Panels are available in lengths up to 16m, but 10m-long panels are used on this project. Panels are lifted into position with a vacuum lifting device (suction pads).

The hydraulic lifter is shown in further images in this section.

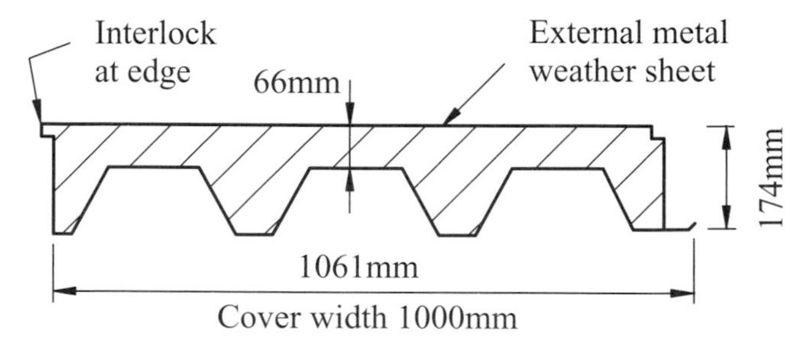

Topspan insulated roof deck

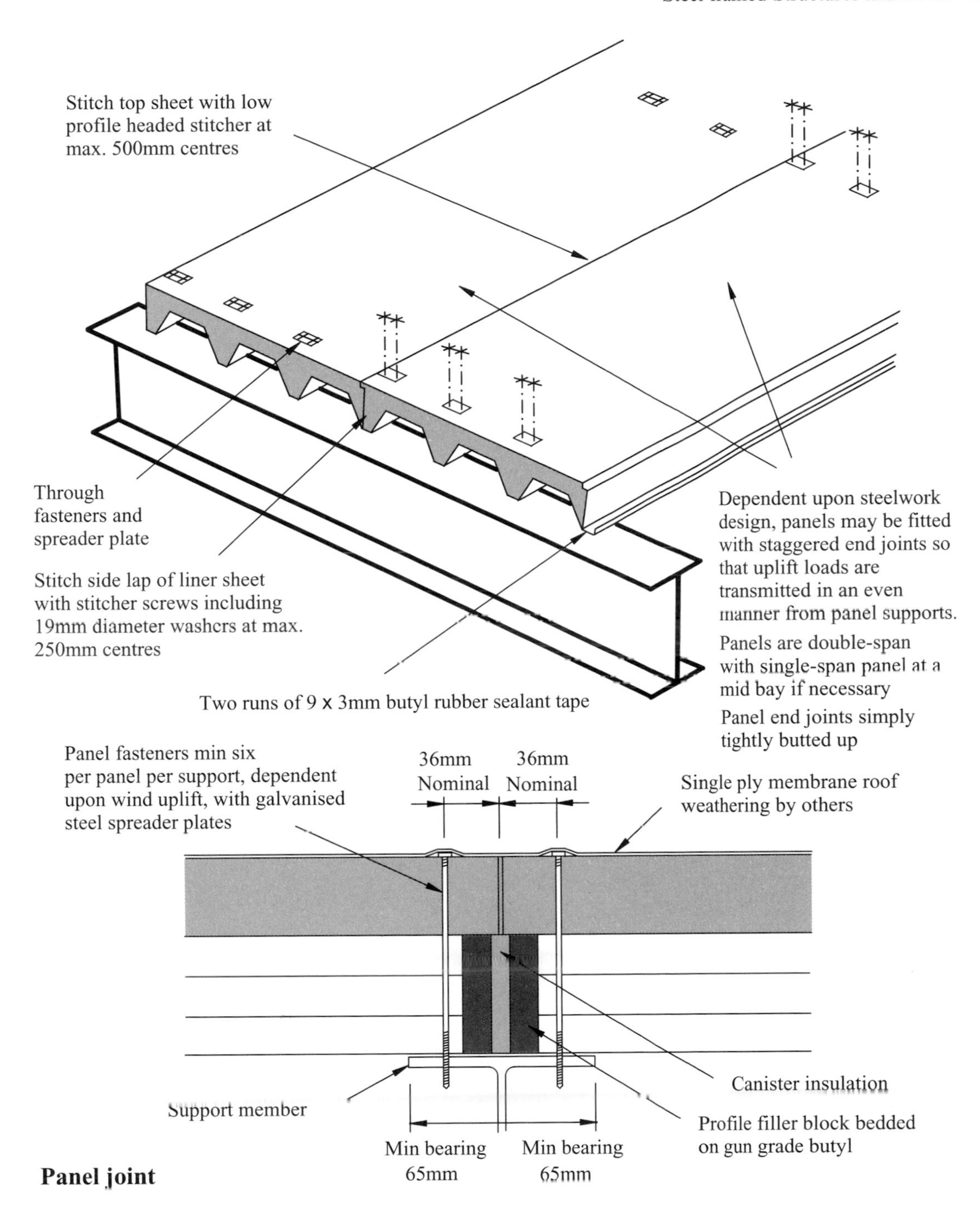
Stitch top sheet with low profile headed stitcher at max. 500mm centres
Through fasteners and spreader plate
Stitch side lap of liner sheet with stitcher screws including 19mm diameter washers at max. 250mm centres
Two runs of 9 x 3mm butyl rubber sealant tape
Dependent upon steelwork design, panels may be fitted with staggered end joints so that uplift loads are transmitted in an even manner from panel supports.
Panels are double-span with single-span panel at a mid bay if necessary
Panel end joints simply tightly butted up
Panel fasteners min six per panel per support, dependent upon wind uplift, with galvanised steel spreader plates
36mm Nominal
36mm Nominal
Single ply membrane roof weathering by others
Canister insulation
Support member
Profile filler block bedded on gun grade butyl
Min bearing 65mm
Min bearing 65mm

Panel joint

Cellular beam frame

Roof decking

Lifting 10m-long insulated roof panels with suction pad hydraulic lifter

Lifting attachment allows the roof panels to be rotate into their final position, i.e. flat surface upwards

Finish in single-ply plastic sheet (Sarnafil product)

Commencement of plastic roof covering to insulated panel roof

Adhesive being applied to roof deck prior to applying roof covering

8.4 Space frames

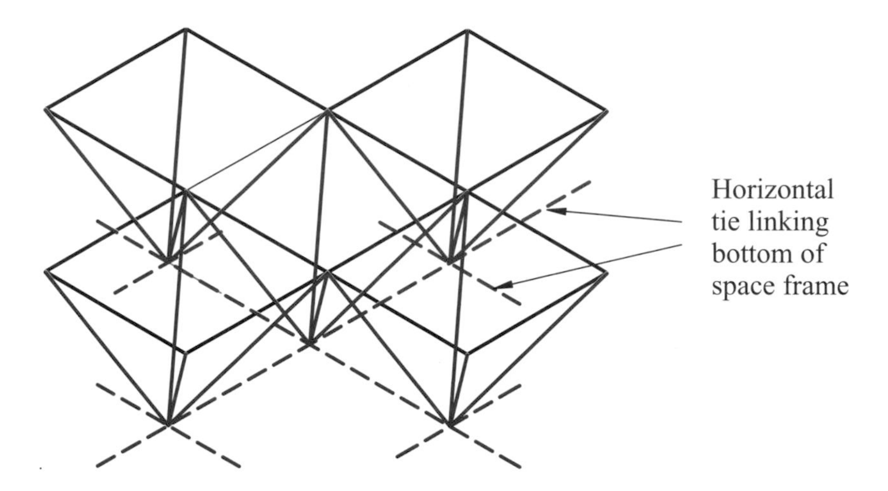

Exploded view of space frame arrangement

Space frame roof canopy

Manchester terminal 2–space frame roof

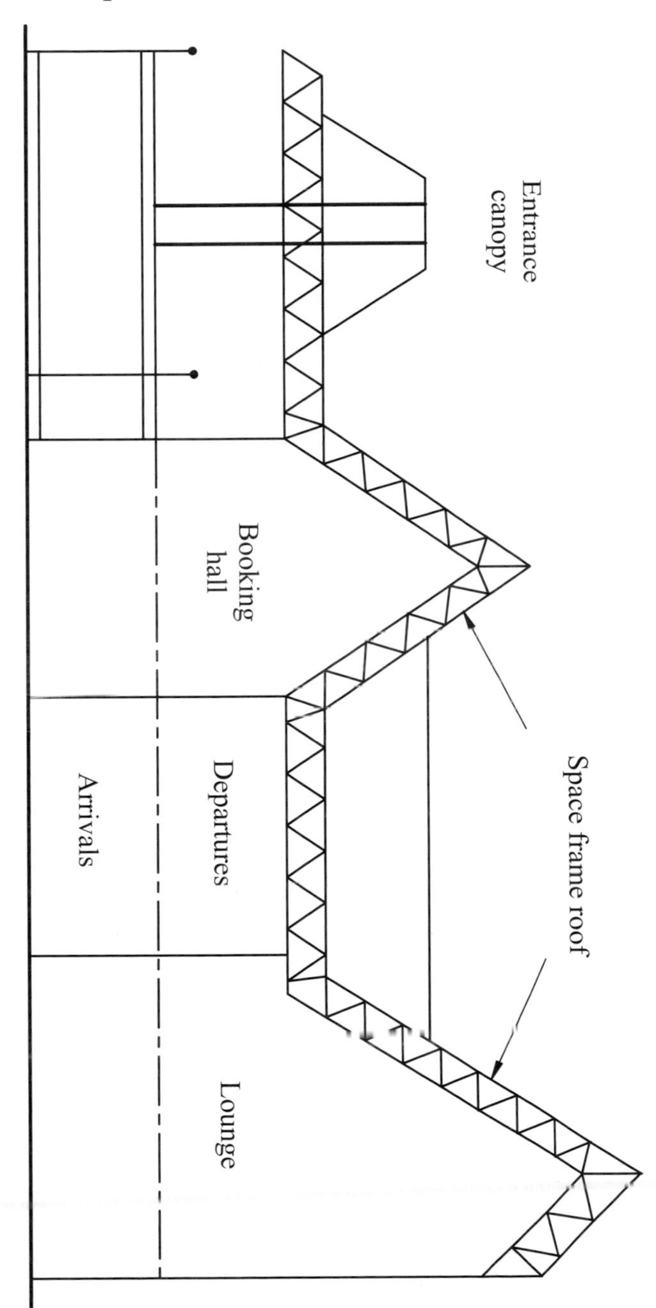

Manchester terminal 2–roof main building

A space frame is a truss-like lightweight rigid structure constructed from interlocking struts in a geometric pattern.
Space frames utilise a multidirectional span and are often used to accomplish long spans with few supports.
(from Wikipedia)

Main space frame roof over departures, booking in and lounge areas
Roof fabricated on the ground and lifted into position in five sections

Manchester terminal 2–entrance canopy

The Terminal 2 space frame canopy extends along the frontage of the airport arrivals building. The canopy is supported on a series of tubular towers.

The canopy roof links into the entrance to the main building.

This project can be readily visited by student groups as it is a public building.

8.5 Arched tubular roof

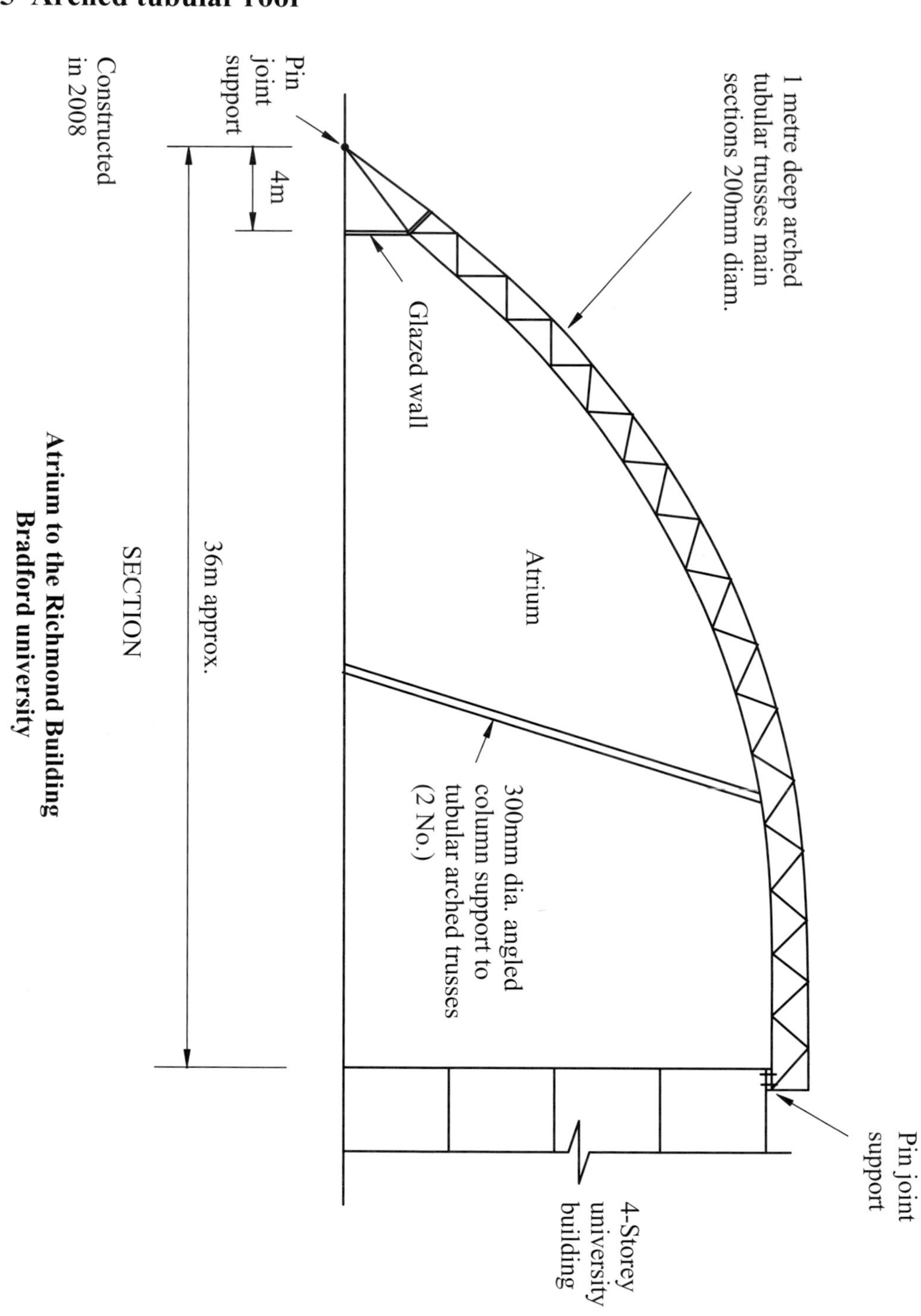

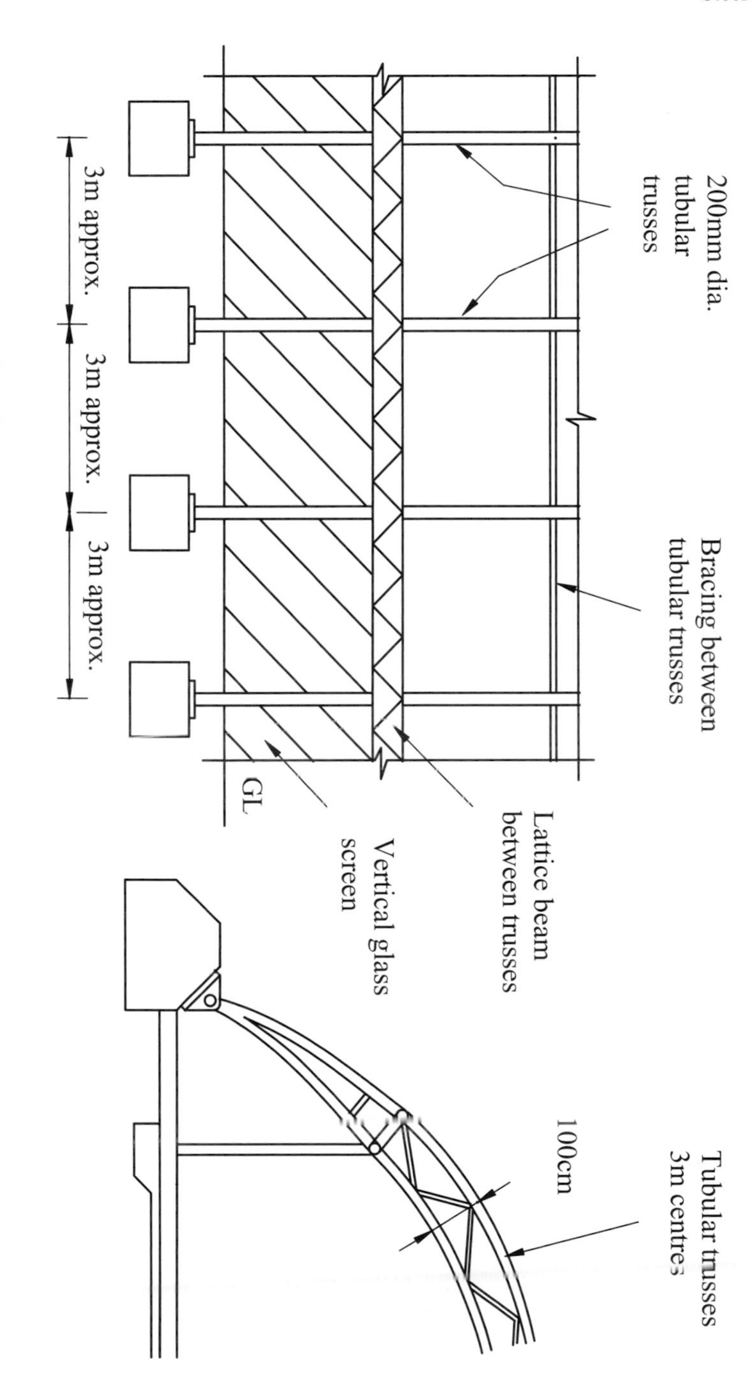
200mm dia.
tubular
trusses
Bracing between
tubular trusses
Lattice beam
between trusses
Vertical glass
screen
GL
3m approx.
3m approx.
3m approx.
Elevation
Tubular trusses
3m centres
100cm
Section

Arched tubular roof

**Richmond Building atrium
pin-jointed tubular arch**

View of atrium showing the tubular arched frame. The twelve arch supports are anchored on a concrete plinth foundation base. The roof is covered in polycarbon sheeting

Internal view of the 35m-wide building with the tubular arched trusses at approx 3 metre centres

Views of external bases

The 200mm-diameter main tubular supports are connected to the foundation base with a stainless steel pinned joint. The joint angles are welded to a 400mm square base plate and fixed with four holding-down bolts.

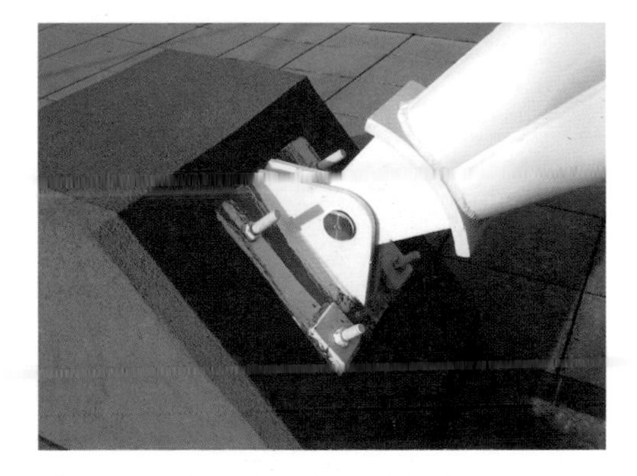

Arched tubular roof

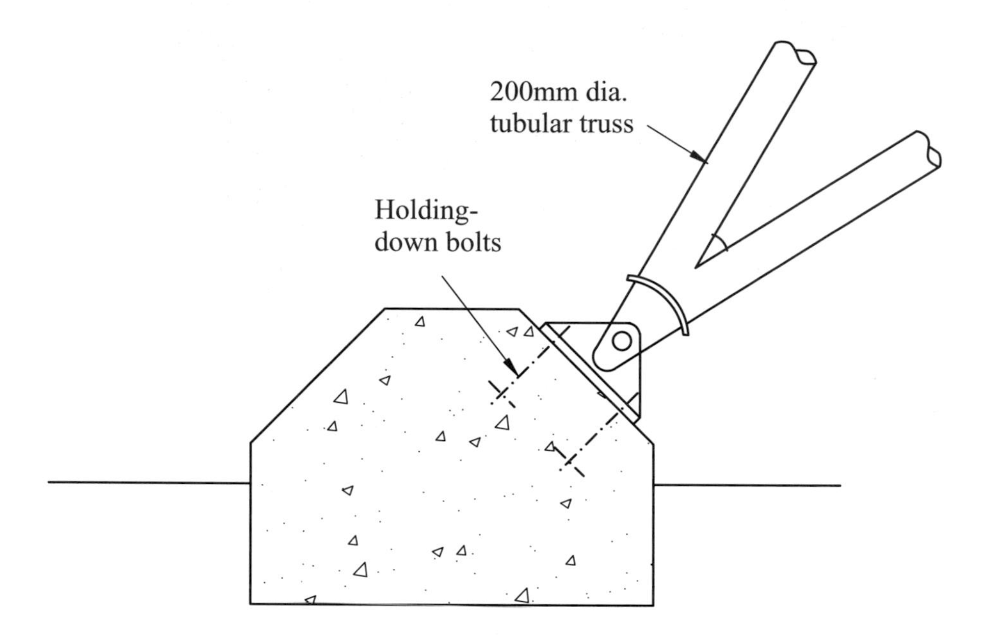

Elevation

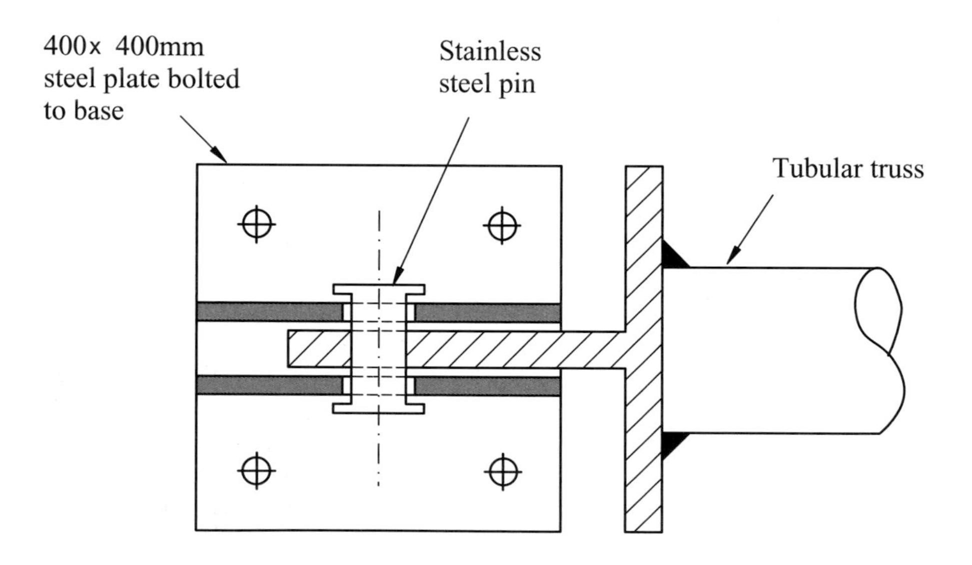

Plan–pin joint

8.6 Cable-stayed buildings

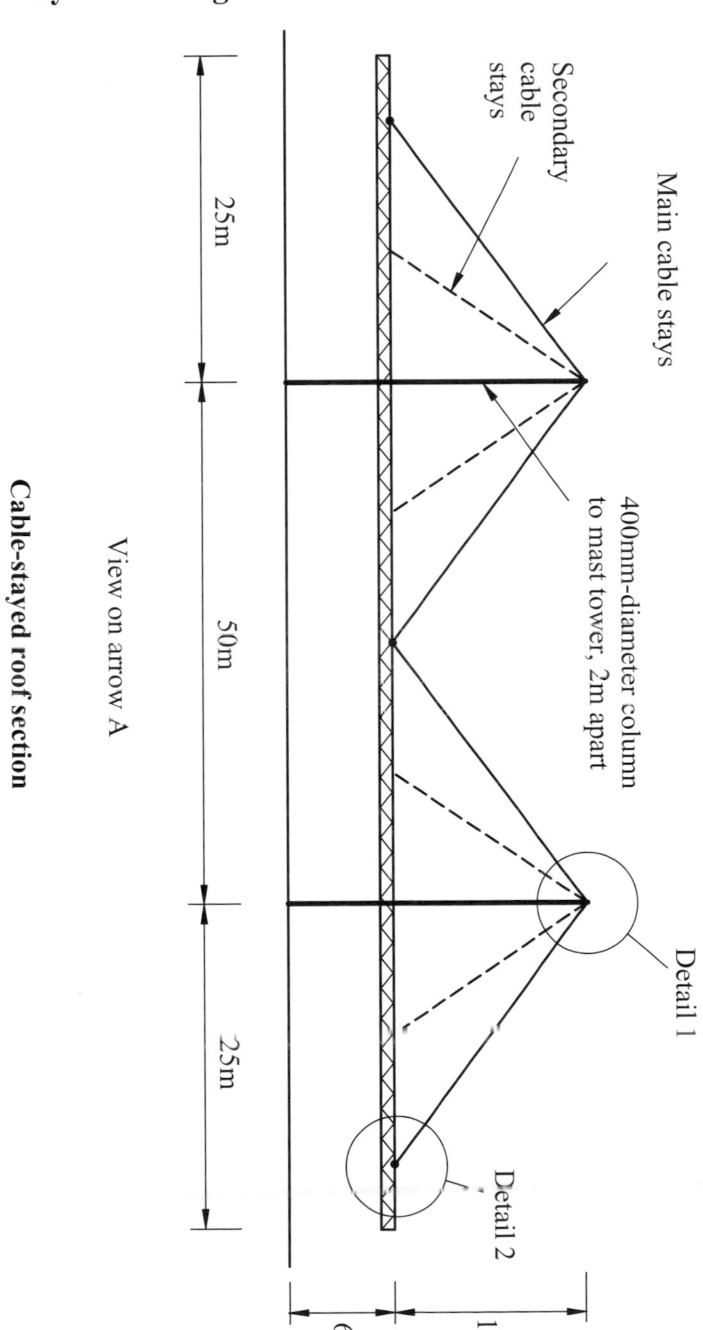

Cable-stayed roof section

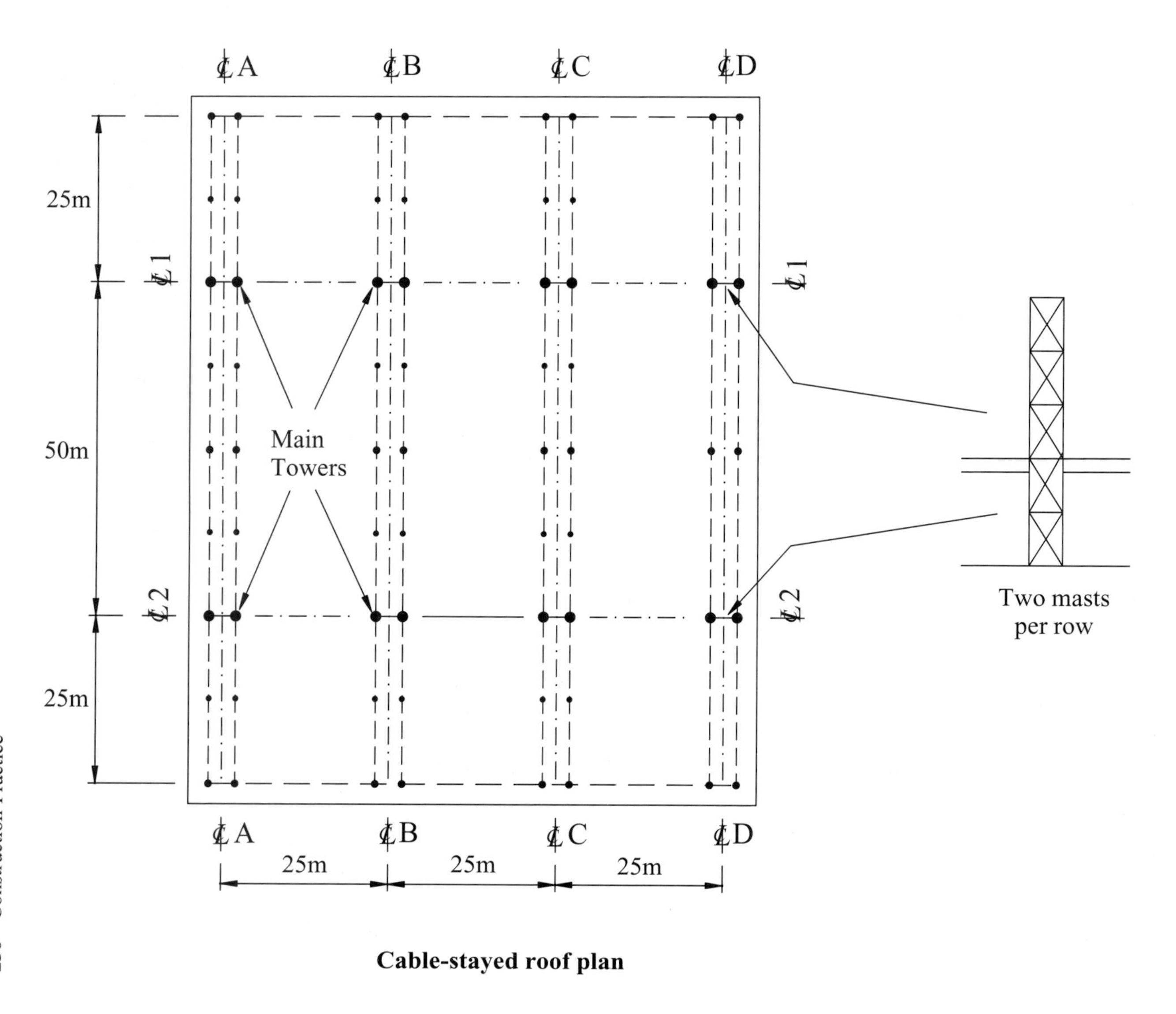

Cable-stayed roof plan

Cable-stayed buildings

View along system line A

The 400mm-diameter steel towers are 10m high above the main roof line. Note the tie bracing to the towers.

View along system line 1 & 2

The four masts can be seen with the main and secondary stays in position.

DETAIL 1

Connection of double main and secondary cable stays at head of steel towers.

DETAIL 2

Connection of cable stays at roof level.

The main roof may be constructed on tubular trusses spanning between steel towers.

Alternatively the whole roof structure may be constructed as a space-frame roof. This project incorporates both of these forms of construction.

8.7 Cable-stayed roof to Metro station

Metro station–Manchester / Oldham link

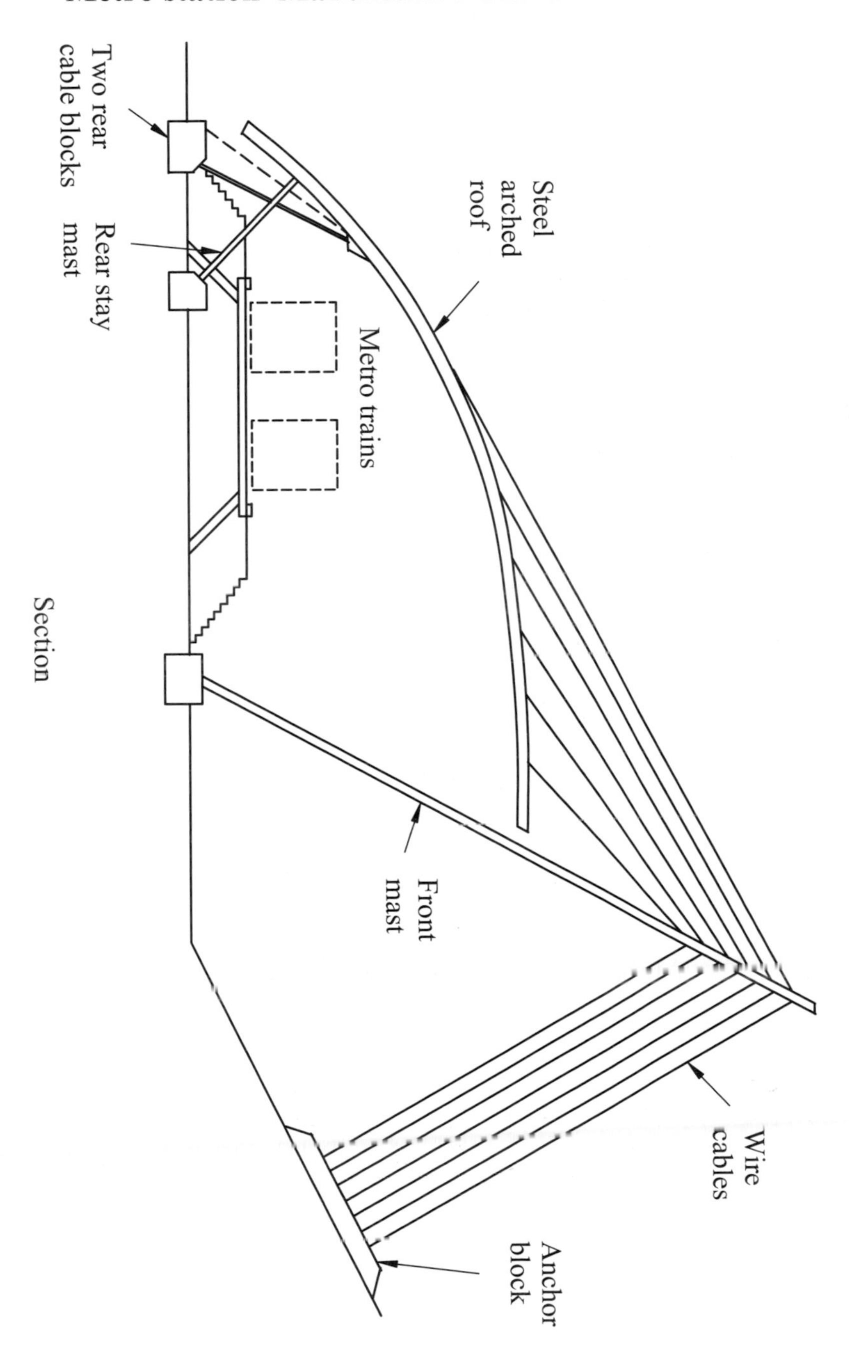

General view of Metro station: steel shell roof supported on cables attached to frame mast cables at front and rear, tied to anchor blocks

View of rear mast stay supporting the roof.
Rear anchor blocks can be seen

Cable-stayed roof to Metro station

Spectacular view of steel canopy, showing the staircase access to the Metro platform area

Rear view of Metro station showing roof cables connecting the roof to the main mast

View of rear cable anchorage

Definition of cable structures

Form of long-span structures that is subject to tension and uses suspension cables for support.

A cable-stayed roof is supported from above by steel cables radiating downwards from masts that rise above the roof level.

(Source: Wikipedia)

View of main cable mast anchorage.

8.8 Bow-string roof to entrance foyer

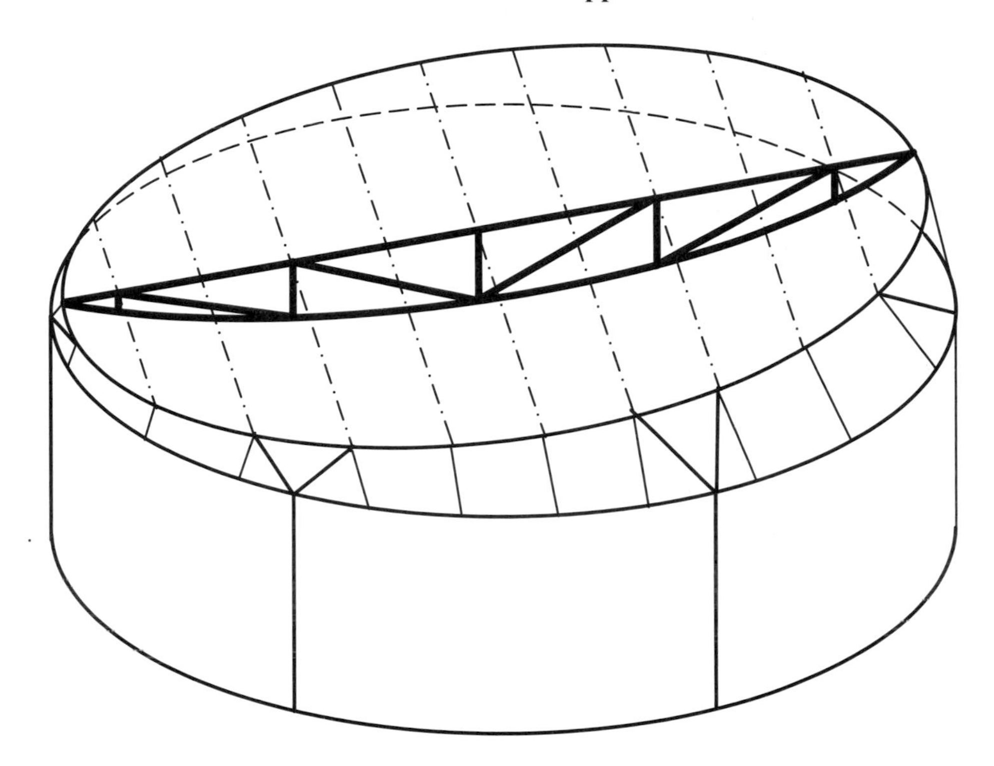

Atrium roof entrance lobby

The isometric drawing shows a bow-string tubular truss supporting the circular tubular roof structure to the atrium of an entrance lobby.

The circular tubular structure is supported on eight 200mm diameter columns. The overall roof is approximately 16m diameter.

This form of roof can be adapted in any location requiring natural lighting, e.g. a car showroom, library building, office area or small exhibition hall.

Photo images illustrate the tubular bow-string truss supporting the tubular purlins. The main glass roof is set to an angle thus creating the interesting glazed panelling below the main circular roof area.

Detailed connections are illustrated of the column supports and the cable tension rods supporting the glass roof panels.

General view of sloping tubular atrium roof structure supported on bow-string tubular truss. Roof structure supported on eight-250mm diameter columns

Tubular circular roof showing bow-string truss and supporting glazing system to glass roof

8.9 Glulam portal frame–church project

Internal roof to church community hall

Internal view of complete church community hall

Detail of portal frame and internal roof layout creating a striking internal feature

Foundation plan

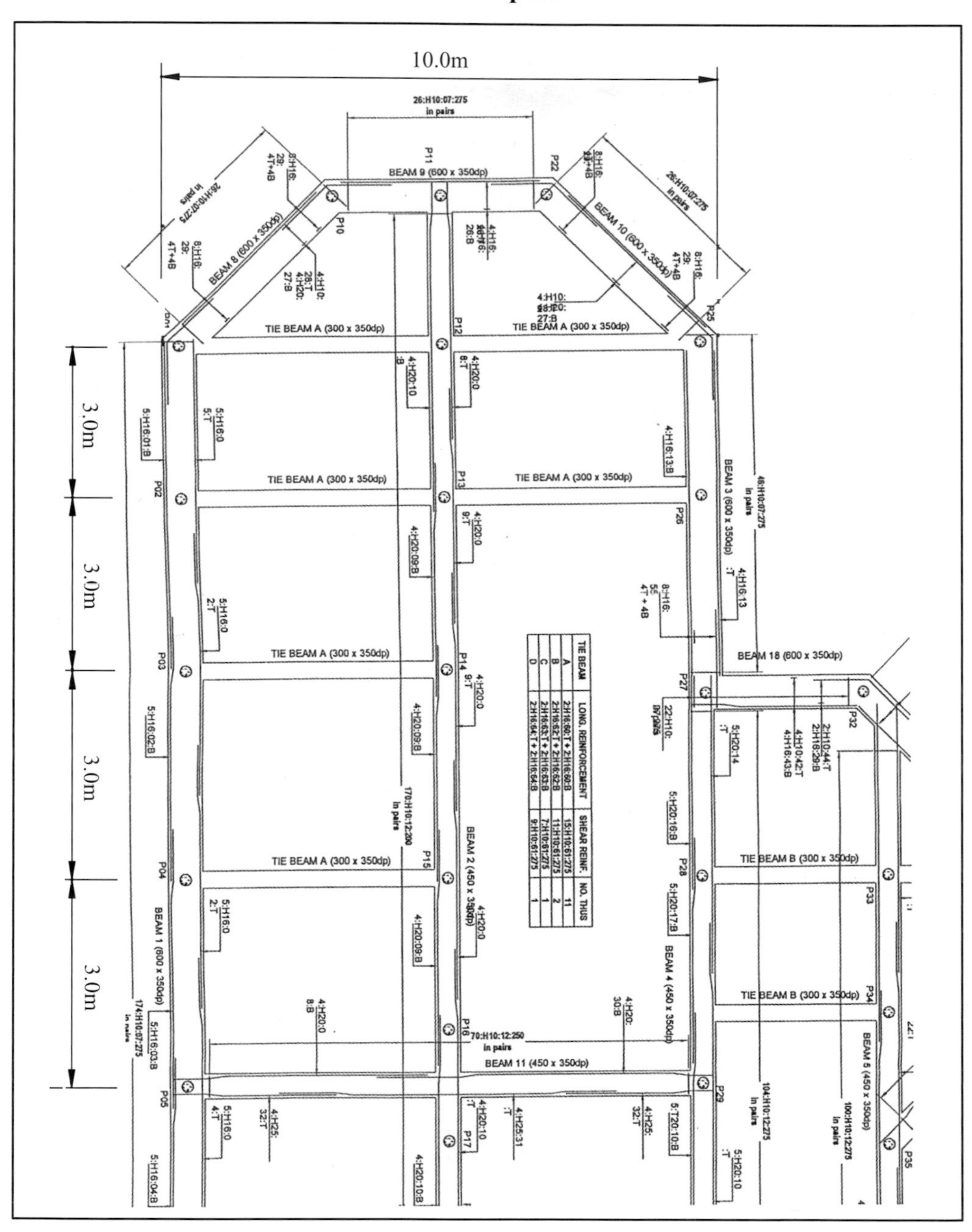

Pile and ground beam layout

Erection sequence

Stage 1
Erect portal frames 1 and 2.

Stage 2
Erect purlins and ridge to connect frames together.

Stage 3
Erect four gable frames GF1 - GF4

The sequence of work involving the erection of the portal frames first includes the erection of access scaffold.

The three stages were undertaken in 8 days. This was followed by the internal timber wall framing between the portal frames and the commencement of the roof boarding. A layout plan of the portal frames and the purlins are shown below.

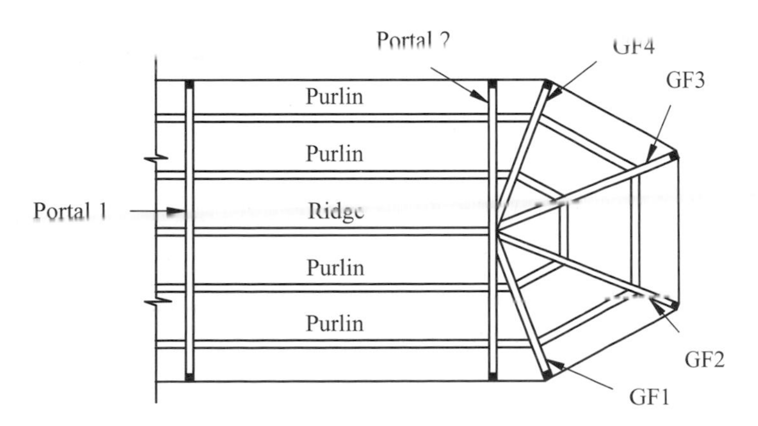

Glulam portal frame

Each portal frame consists of two sections jointed at the apex. The 430 x 215mm column section is factory-fitted with a galvanised shoe. The shoe is bolted to the in-situ floor slab using four resin-drilled anchor bolts. Details of the shoe fixing are shown.

Internal wall construction consists of site-constructed timber stud panels fitted between the portal frame support columns. On the inner face of the cavity they are plywood lined.

The internal timber framing is insulated with rockwool bats. Internally, the building is lined in vertical softwood boarding up to dado level with a plasterboard lining and skim to eaves level. Externally, facing brickwork forms the building enclosure.

Roof construction

Galvanised connectors are used to connect the ridge and purlins to the main portal frames.
The connections are all site fixings. The purlins and ridge are then positioned in galvanised slots and further secured by nailing.

The prefabricated roof frame for the roof structure over the building entrance is shown. This was fabricated on an adjacent scaffold platform and craned into its final position prior to applying the roof covering.

Completed roof tiling

Structural decking being fixed

70mm insulation layer in position

The roof is finished in dark clay tiles.

The build up of the roof construction is as follows:

① 32 x 125mm redwood t and g boarding structural decking

② 1200g visqueen barrier

③ 38 x 63mm softwood battens

④ 40mm Kingspan insulation

⑤ 80 x 38mm softwood rafters

⑥ 70mm Kingspan plyurethene insulation between battens

⑦ 20 x 50mm tanalised roof battens on breathable sarking felt

⑧ Concrete tile finish

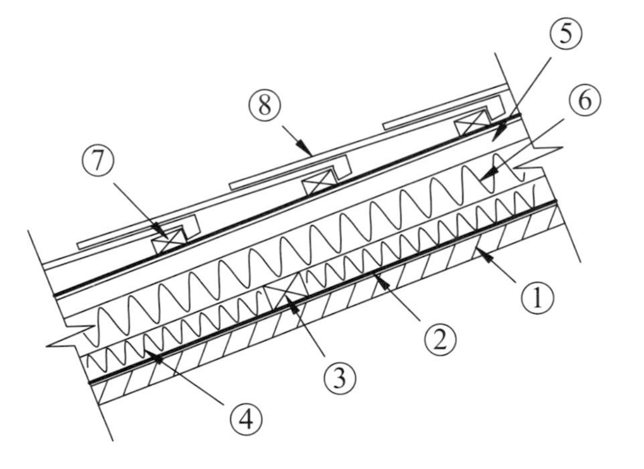

Roof section

8.10 Glulam portal frame–Sheffield winter gardens project

The Winter gardens building is approximately 70m in length, 22m wide and 21m high. It consists of ten main parabolic Glulam frames constructed of larch timber. The frame requires no preservation or treatment.

The main portals are pin-jointed at the base of each frame supported with a metal shoe. Horizontal and diagonal bracing is provided between the portal bays. The building is one of the largest Glulam laminated frames in the UK. The building forms part of a £120m regeneration project. The complex incorporates the Sheffield museum and the Mecure hotel. The building was opened to the public in 2003.

Horizontal and diagonal bracing to main Glulam arches.
Main parabolic arches at approx. 9m centres.

Pinned joint at frame support

Architect–Pringle Richards Sharratt
Structural Engineer–Buro Happold
Services Engineer–Buro Happold
Quantity Surveyors–Sheffield City Council

The contract was based on a management contract undertaken by Interserve Project Services Ltd.

The Wikipedia web site is a free encyclopedia site which allows access to a wide range of building projects throughout the UK and the world.
Ref: www.wikipedia.org/wki/list-of-buildings and structures (why not try it)

CHAPTER NINE

FORMWORK IN CONSTRUCTION

Construction Practice. Brian Cooke

9.0 Overview

A range of formwork applications are illustrated aided by related sketches.

- Formwork to a freestanding wall
- Formwork to a basement retaining wall using patent formwork
- A six-storey lift shaft, using the Doka system
  The storey-height lifts being raised by a mobile crane
- Various types of column shutters including metal shutters, traditional column cramps and PVC circular column formwork
- Formwork systems to floor slabs, using patent floor props
- Case studies related to the application of sliding formwork for staircase tower cores–the method of constructing these cores is difficult to understnd. Two projects in the north-west: Seven-storey staircase core at Castlefield and 16-storey twin cores at Mann Island, Liverpool

  Access to these project was kindly given by the main contractor.

Using photo images of construction allows one to clearly illustrate how the erection process works.

The formwork examples should help students to understand temporary works processes more clearly.

Setting up support formwork for flat slab construction

9.1 Formwork to foundations

Permanent formwork support systems to foundations

Usc of PECOFIL permanent formwork (by BRC) to a raft foundation
Central area to be stone backfilled prior to concreting

Polystyrene soffit and formwork support (Dutch)

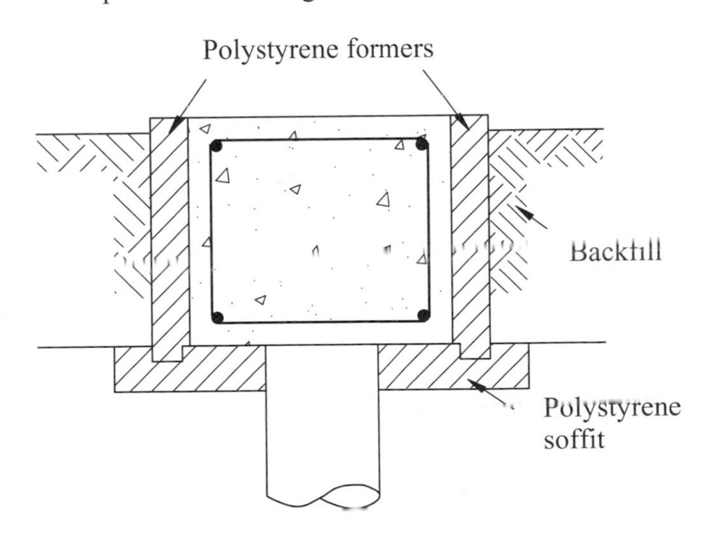

Permanent former left in position

9.2 Formwork to walls

Formwork to 5m high wall

Formwork is used for a 5 metre high free-standing wall. Aluminium wallings and vertical strongbacks have been used for the formwork system. An access platform is provided to assist the placing of concrete.

Full-height metal shutters with access platforms on both sides of the wall are available for crosswall construction.

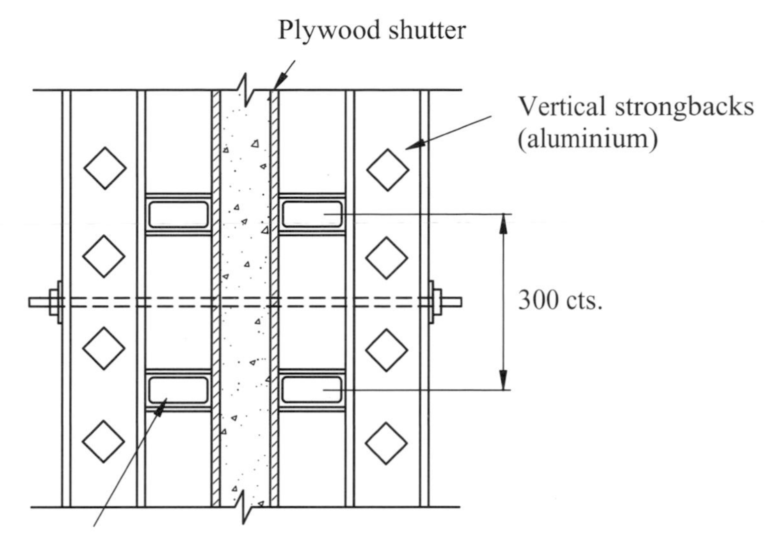

Detailed section through wall formwork

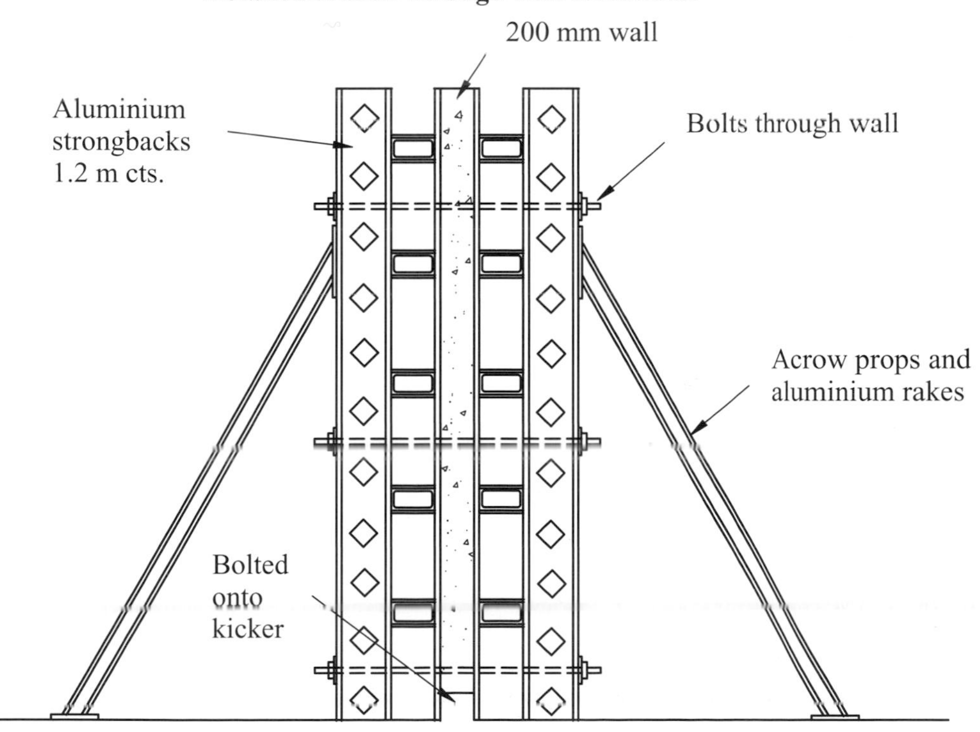

Wall formwork

Two applications of formwork to wall situations are illustrated.

Formwork is shown to a 2m-high basement wall using patent aluminium strong-backs with additional propped support using Acrow props. Many types of form systems are available to hire or buy including Mabey, RMD and many others. A list of reference web sites is given in the web appendix.

On the continent, cross wall construction is extensively used for residential housing. The crosswalls are constructed 2.5– 2.8 metres in height across the whole width of the building. Full-height metal shutters are illustrated with access platforms located on each side of the wall.

Six-storey lift shaft–storey-height formwork

Two six-storey lift shafts are under construction using 3m-high formwork lifts. The Doka formwork system uses patent aluminium shutters, formwork raised to the next lift by mobile crane (see later images). The wall pours are concreted using a mobile crane and skips.

Doka formwork system being used for the full-height lifts on the six-storey lift core. Wall pours undertaken with mobile crane skips.

Self climbing formwork systems are available from Peri formwork systems.

Six-storey lift shaft

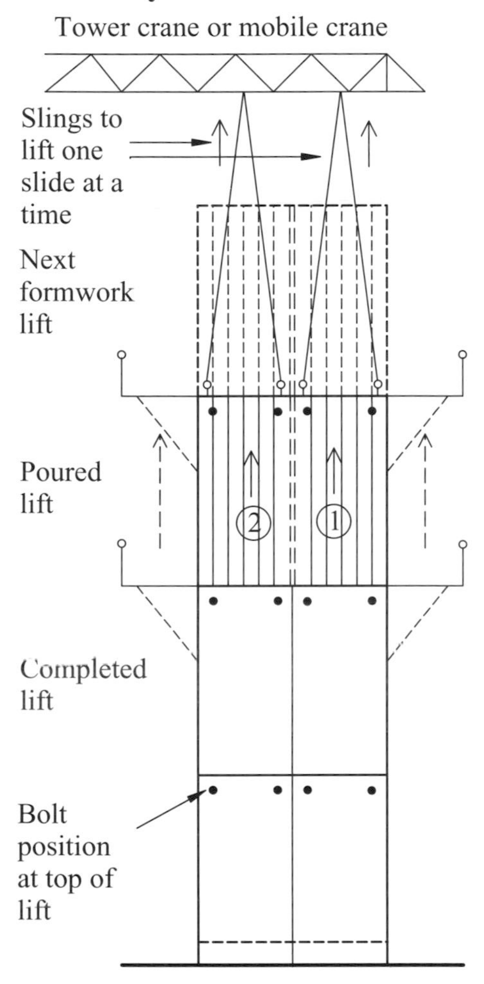

Platform and attached formwork lifted one floor at a time by mobile crane (lifted in two halves).

Internal wall formwork positioned first followed by external platform and attached shutter

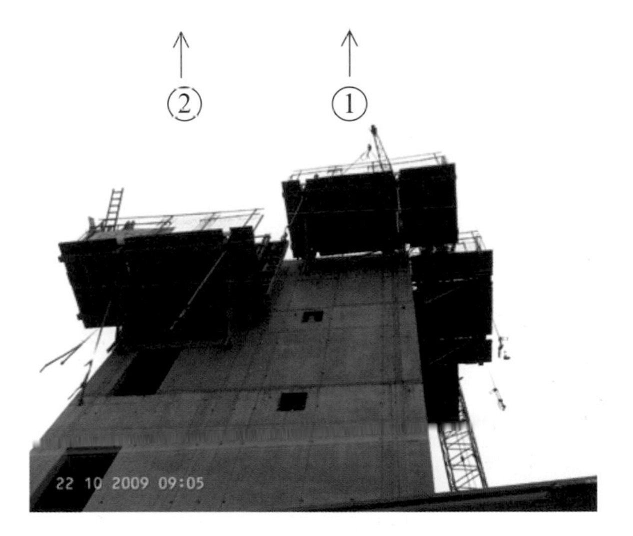

Formwork raised to next position and bolted to top of completed pour

9.3 Formwork to columns

A range of column formwork in use on a single contract

Steel column shutters with aluminium props

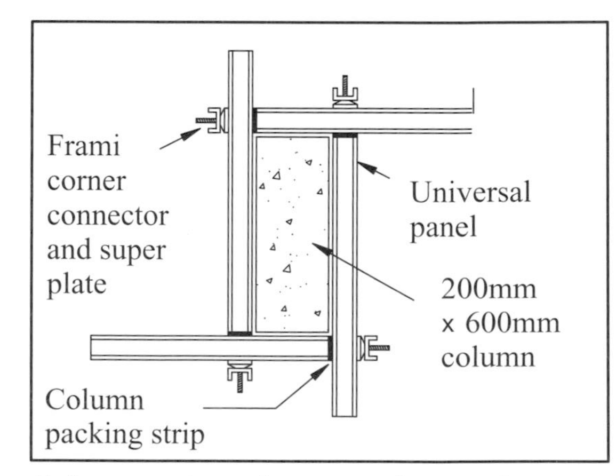

Column section using patent formwork system

Traditional column formwork with column cramps

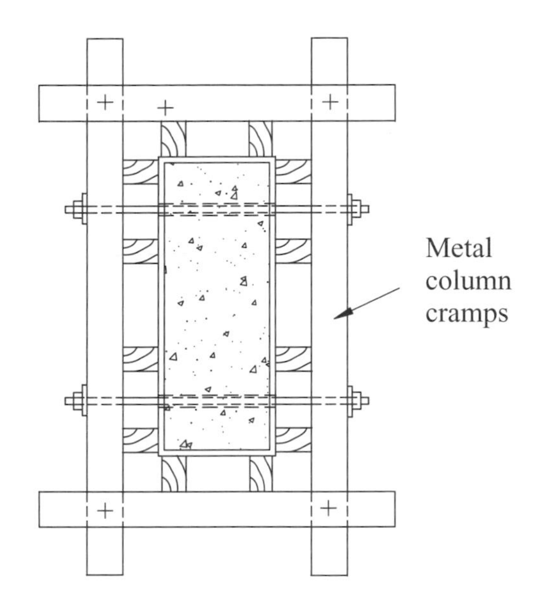

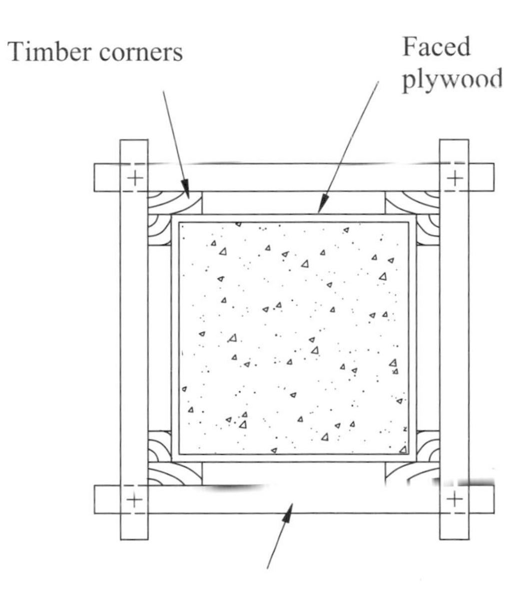

300 x 300 mm
Column shutter

A modern PVC column former to circular entrance columns

9.4 Formwork to floor slabs

Commencement of formwork support to floor

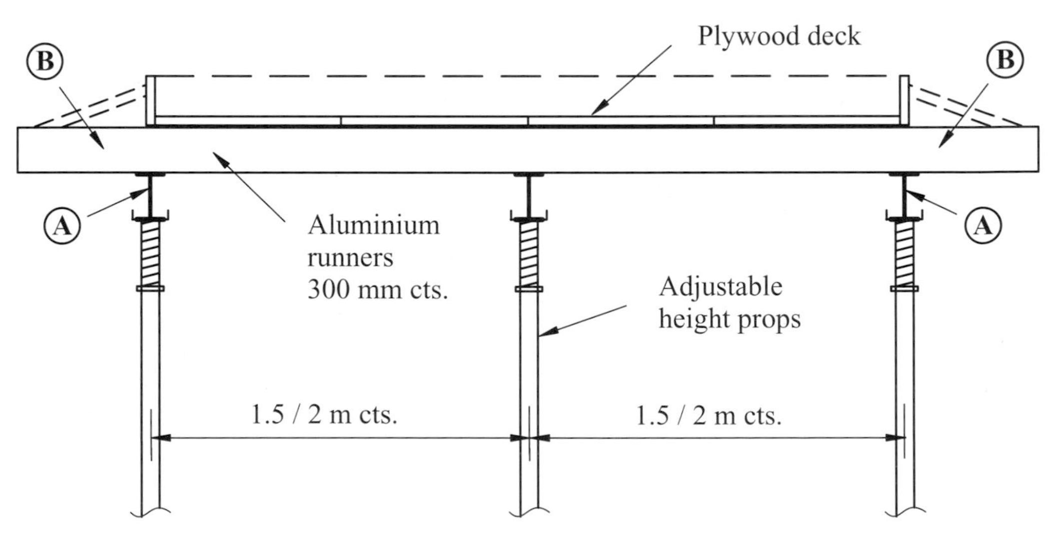

Section through formwork

Formwork system to suspended floor slab.
(similar formwork system to previous page)

Examples of Peri Skydeck formwork used for slab soffit support
The system is all aluminium and the panels are small and easy to handle

9.5 Sliding formwork for staircase tower cores

Sliding formwork is a fast and economical construction method and can be used on numerous types of projects such as stairwells, lift shafts, walls and silos.

The technique is based on movable steel formwork moulds that are lifted by hydraulic jacks. The jacks climb up jacking bars which are built into the concrete wall as work proceeds

From a suspended scaffolding directly below the mould, the concrete surface that appears under the mould can be brushed before the concrete hardens.

Benefits of sliding formwork

- rapid installation time
- shorter construction period
- reduced overall costs
- no casting joints
- opening and embedded items readily accommodated
- safe construction method
- system unaffected by weather

Assembling the sliding formwork at foundation level

Bracing to core area of formwork

First lift pour in progress

Steel yokes shown on concrete pouring platform

View of 50mm diameter jacking bars used for raising the platform

Sequence of work during slide

The formwork is raised 3m per day, with concrete placed as the shutter is moved vertically upwards via the jacking bars.

The number of men operating on the working platform is:

Concreting and jacking operations	4 men
Steel fixers	2 men
Labourers–rubbing up "green concrete"	2 men
Joiners–fixing slots and openings	2 men
	10 men

The amount of concrete in a 3m slide at Mann Island is about 30c.m.

Sliding formwork detail

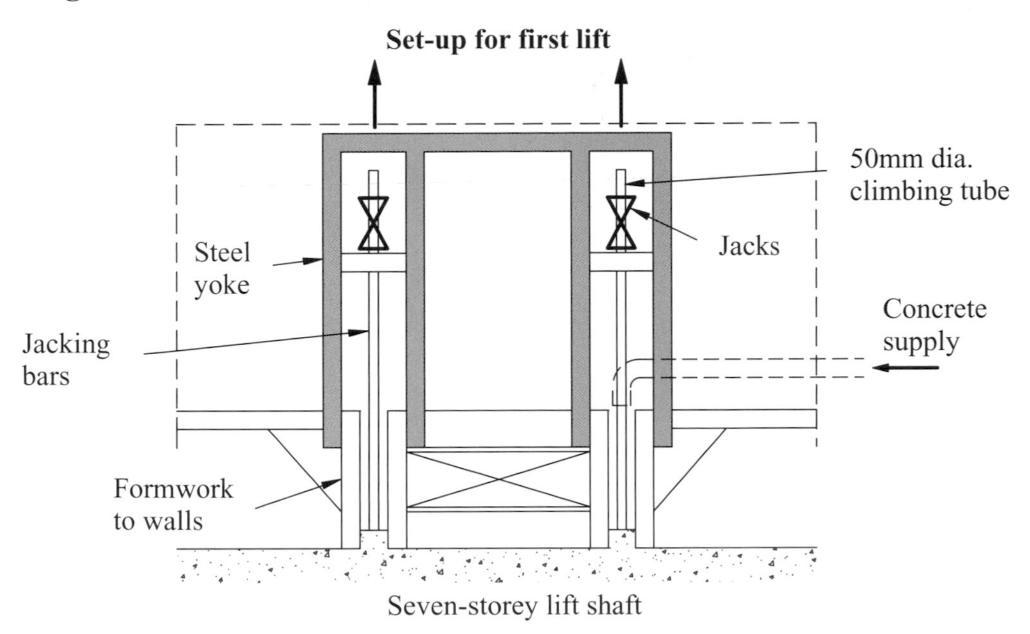

Seven-storey lift shaft

Pouring access to walls
using crane and skips (trcmi pipe)

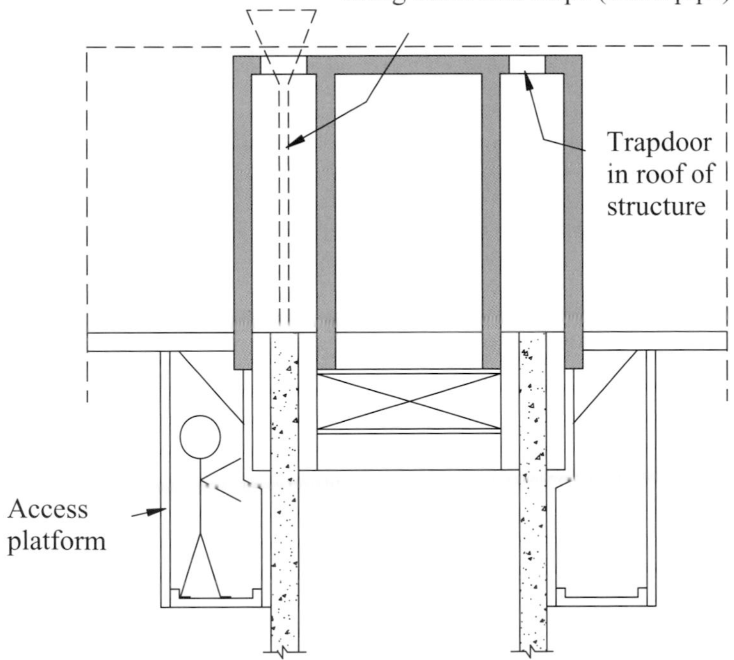

Access for checking quality of completed wall to higher lifts

What is sliding formwork? Typical description

The formwork or mould height is 1.1m and constructed of metal or timber formers covered with a 1mm thick steel sheet. Lifting yokes and hydraulic jacks are fitted to the scaffolding that is generally moved upwards as the slide progresses. Lasers are used to check the vertical structure. Horizontal (level checks) relate to datum levels marked on the structure.

The work platform is fitted to beams with a hanging platform mounted below. During operation all jacks are contolled from a central unit.

Each lift normally raises the platform 20–30mm (one inch). Casting and reinforcement work are normally performed continuously as the sliding formwork moves upwards.

The intended hardening front is 900mm. This is located about 200mm from the bottom edge of the formwork mould.

The setting time of the concrete is regulated by additives (accelerants or retarders) so that it matches the progress of the mould.

A variety of formwork systems are available including the Slipform system. This is being used for the twin towers and single tower construction at the Mann Island project on Liverpool's waterfront. The work is being undertaken by PC Harrington (Slipform UK) and progresses at the rate of 3 metres per day. The system is based on raising steel yokes on jack bars incorporated in the wall construction.

Other systems include the use of self climbing-formwork available from Peri Ltd (formwork scaffolding engineering). See web sites listed in the appendix.

Liverpool's 'Twin Towers'

Life on the deck platform

The steel yokes and jack bars

16 floors in 16 days

CHAPTER TEN

FLOOR CONSTRUCTION

Construction Practice. Brian Cooke

10.0 Overview

A range of forms of floor construction are illustrated for both domestic and commercial building projects. Photo images have been carefully matched with a range of construction sketches.

- **Beam and infill block floors**

  These are often referred to as 'beam and pot' floors. The prestressed I beams may be filled with solid or hollow blocks. The blocks may be manufactured of concrete, airated concrete or lightweight aggregates. Polystyrene formers may also be used which provide for an insulated floor.

  Examples illustrate the use of polystyrene formers for a residential development in Derby. A further example shows a heavy beam and block floor for a small housing contract.

- **Metal deck floor**

  Metal deck floors are incorporated into muti-storey steel-framed buildings. The metal deck is shot fire fixed to the top flange of floor steelwork.

  Shear bolt fixings are also incorporated to tie the floor and slab to the frame and provide lateral stability. A 100–150mm in-situ concrete slab is usually poured to form the structural floor.

  As an alternative to a metal deck, wideslab solid or hollow core precast concrete units may be specified. Both floor types allow instant access to the floor area.

  The construction detail illustrates a standard metal deck floor and a raised floor with an exposed grid suspended ceiling below.

- **Precast concrete floors**

  A wide range of solid and hollow precast floors are available from a wide range of manufacturers. The advantages of precast floors are outlined.

  The application of wideslab precast floors is illustrated for a roof deck to a storage tank. A further set of images relate to a ten storey precast crosswall frame.

- **Precast plank floors**

  Precast plank floors act as a permanent replacement to a formwork soffit on a suspended floor.

  Omnia planks have been available since the 1960s where they were incorporated in the original Jesperson precast crosswall systems. The planks are 50–100mm in thickness depending upon their span.

  They are widely used in Holland, Denmark and Germany for domestic housing and crosswall construction.

10.1 Beam and block floors

Precast beams with polystyrene formers

Beam and block floors

Polystyrene floor formers in lieu of blocks have been widely utilised on the continent for a number of years, especially in Holland and France.

The illustrations show polystyrene formers being used for the ground floor construction on a low-rise housing project.

The floor surface is finished with a DPM and fibre-reinforced concrete screed 80mm in thickness.

Beam and polystyrene block floors

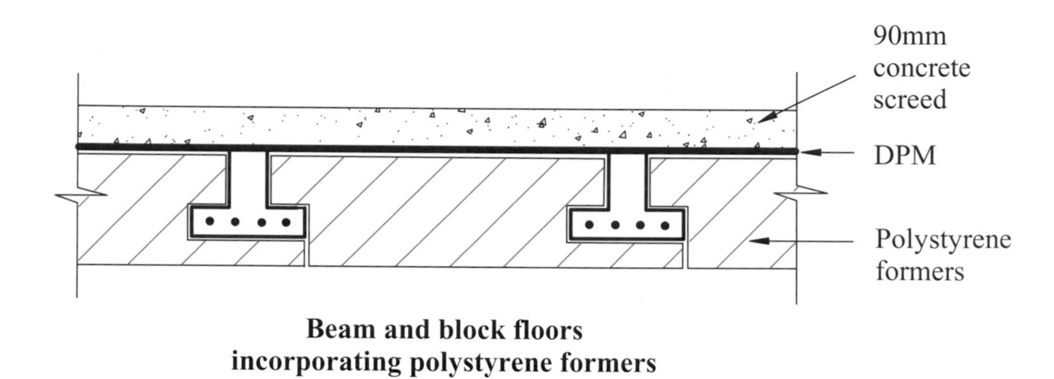

Beam and block floors incorporating polystyrene formers

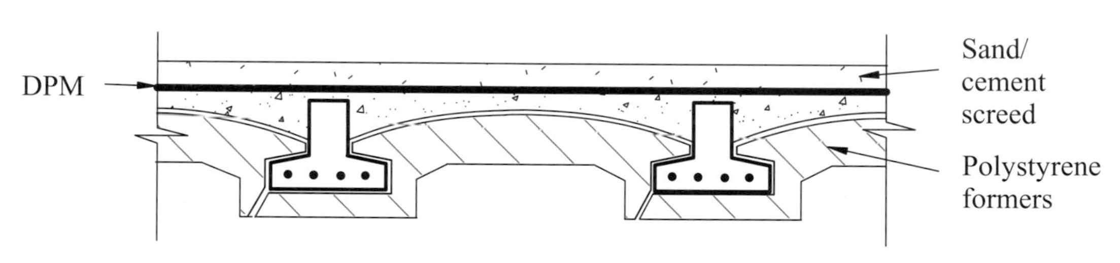

I Beams and polystyrene formers (Dutch)

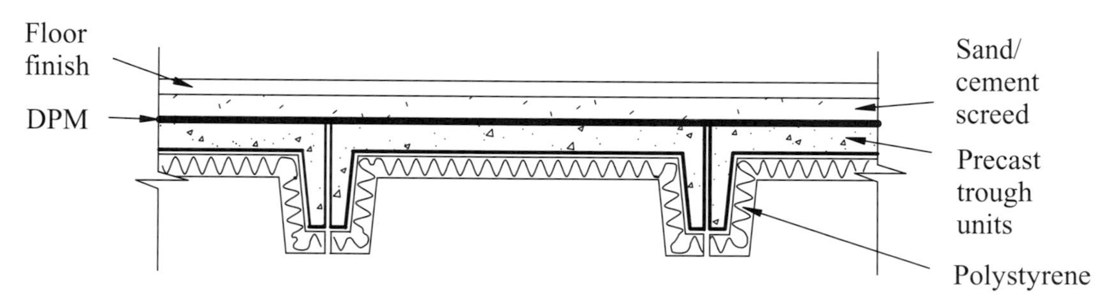

Precast trough units with polystyrene cast on at manufacturing stage (Dutch)

Precast beams with infill blocks

Beam and block floors

A wide variety of beam and block floor types are available from a range of manufacturers.

These include Tarmac Topblock, Hanson Floors, Bison Concrete Floors etc.

The depth of the prestressed floor beams varies from 100mm to 200mm depending on the span and loading conditions.

The illustrations show this floor type being used for ground and first-floor construction.

A ground floor slab to a basement is shown with a short span of 3.5 metres.

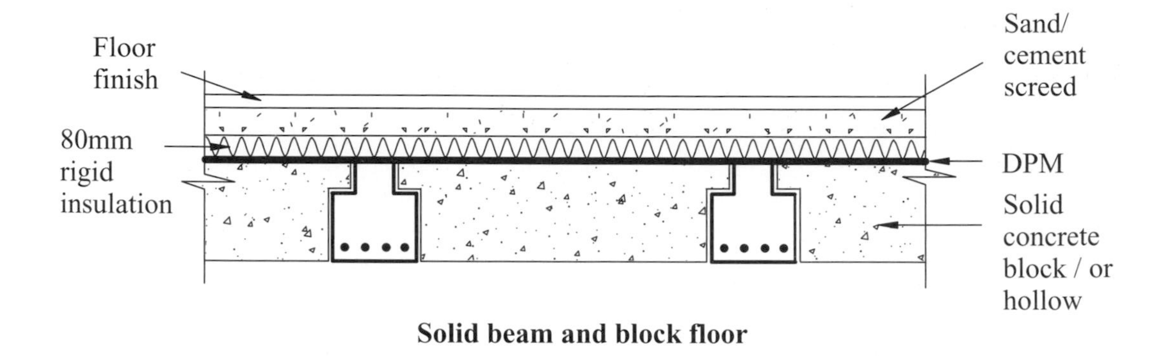

Solid beam and block floor

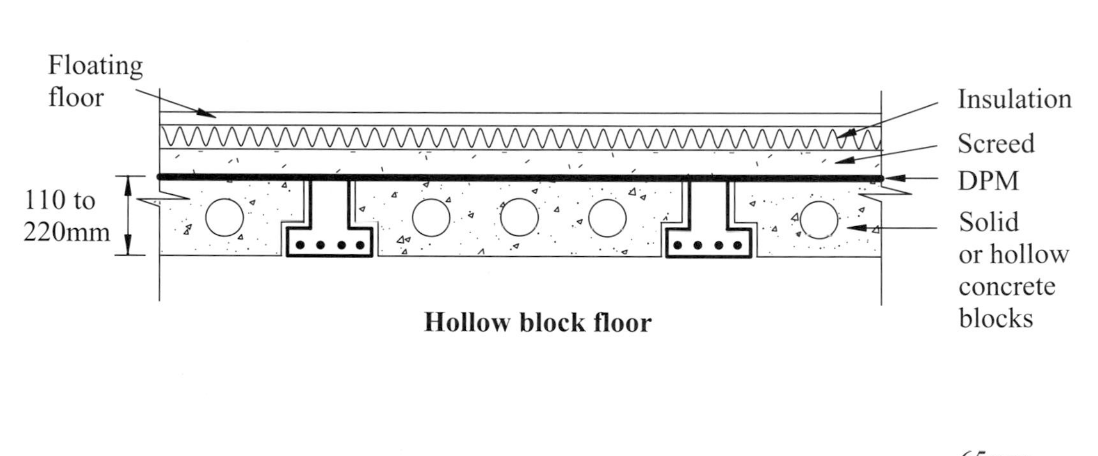

Hollow block floor

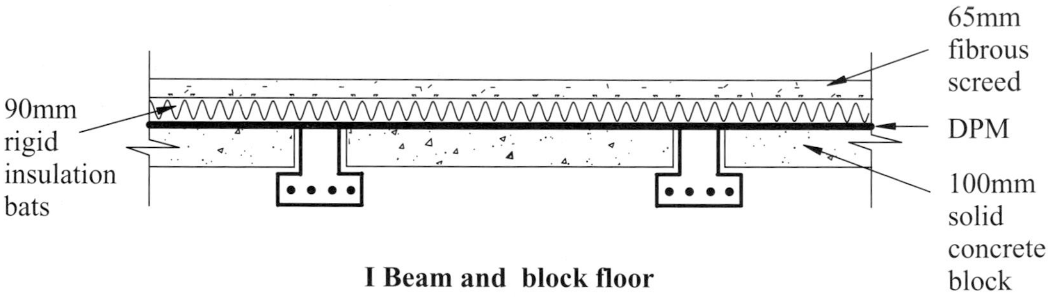

I Beam and block floor

Beam and block ground floor to a five-storey block of residential flats

General view of completed beam and block floor

DPM membrane laid and 90mm-thick polystyrene insulation ready for fixing. Fibrous screed 65mm thick laid over insulation

10.2 Metal deck floors

Holorib metal decking is available in various thicknesses depending upon the span between beams.

The decking is shot fired to the supporting steelwork to provide lateral stability.

Shear bolts are fitted to tie the floor deck to the main frame.

Fabric reinforcement is positioned and the floor concreted 100–150mm in thickness.

The surface of the slab is normally power float finished.

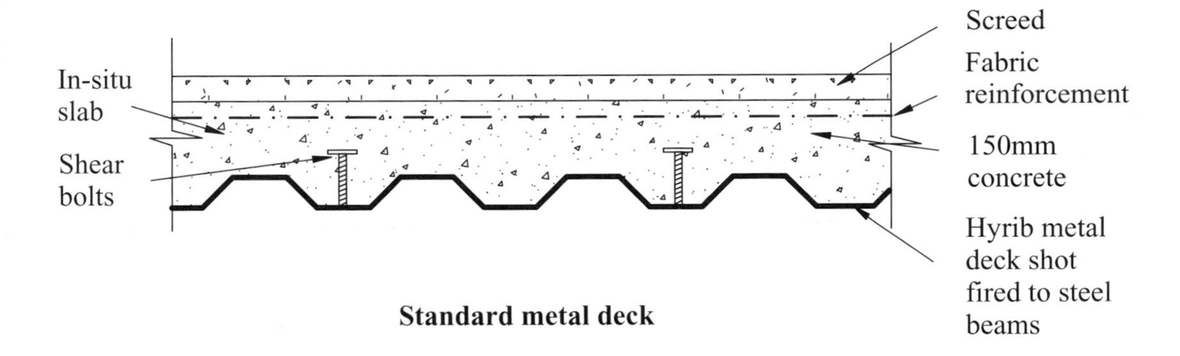

Standard metal deck

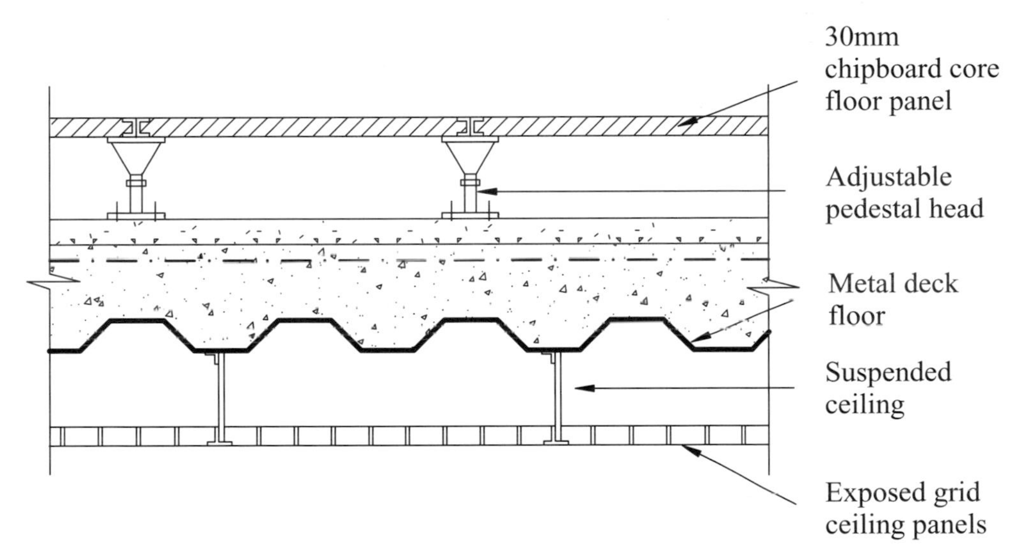

Section through floor with suspended ceiling and raised floor

10.3 Precast concrete floors

Precast concrete floors are available from a wide range of manufacturers. Extensive construction detail is illustrated in their catalogues and data sheets.

Solid floor units	Standard solid floor units 100–400mm in thickness Widths available 600–1200mm Solid composite floor units 125–300mm and 600–1200mm wide Structural concrete topping applied 50–75mm thickness
Hollow core units	Standard hollow floor units 100–400mm in thickness Widths available 600–1200mm Insulated ground floor units available similar sizes with 100mm expanded polystyrene bonded to underside of units Composite hollow floor units 1200mm wide with 50–75mm structural topping

Composite floor types incorporate a structural in-situ concrete topping in order that the prestressed floor units and structural screed act together to form the structural floor.

Precast manufacturers include:

Bison Concrete Products
Hanson Concrete
Longley Floors
Thomas Armstrong

Advantages of precast floors

- Factory produced units ensure high quality control
- Provide fire resistance between dwellings and meet Building Regulations requirements on the passage of fire
- Preformed service holes may be accommodated at the manufacturing stage
- Precast floors form an extremely efficient sound barrier due to their weight
- Speed of erection–forms an instant platform or deck for access and storage
- Design flexibility and structural efficiency
- Wide range of manufacturers leads to a competitive pricing environment
- Good technical advice available

Wideslab floors

Available as solid core units. Floor units are 1200mm wide with thickness 150–300mm.

The basement roof has a 100mm thick structural composite in-situ concrete screed over the units.

Standard hollow core units are available 150–300mm thick. Floor units are also available with polystyrene cores to improve the insulation value (three or five core units are available).

Refer to flooring websites in the appendix.

Tapering precast floor units spanning between crosswalls on 10– storey- block of residential flats

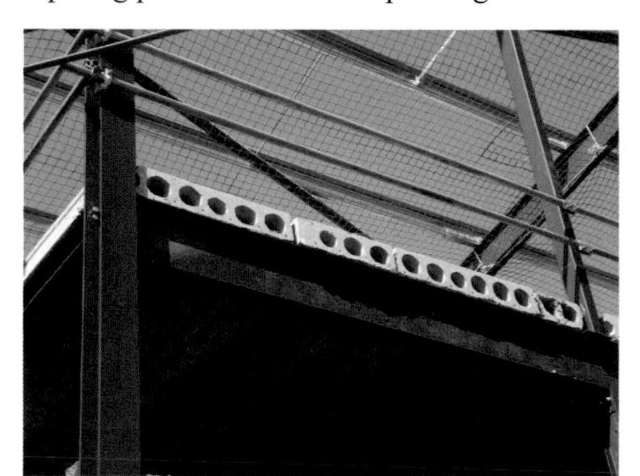

Wideslab precast floor units to first floor of a steel-framed building
Precast stairs positioned on landing

Precast concrete wideslabs

An extensive range of precast wideslab units are available from main manufacturers.

Spans are available up to 12 metres between supporting walls and beams. Continental manufacturers incorporate floor insulation at the manufacturing stage.

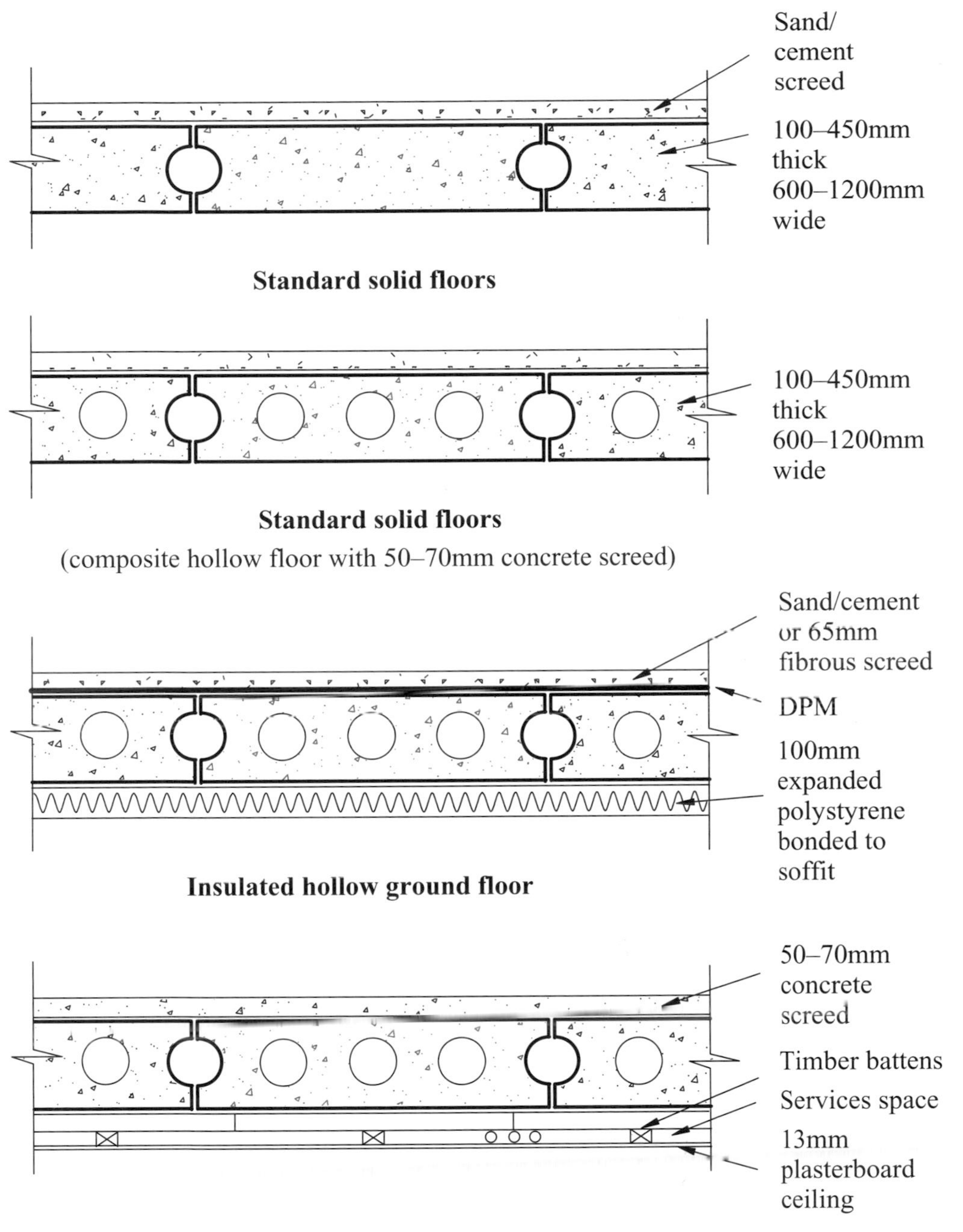
Sand/
cement
screed
100–450mm
thick
600–1200mm
wide
Standard solid floors
100–450mm
thick
600–1200mm
wide
Standard solid floors
(composite hollow floor with 50–70mm concrete screed)
Sand/cement
or 65mm
fibrous screed
DPM
100mm
expanded
polystyrene
bonded to
soffit
Insulated hollow ground floor
50–70mm
concrete
screed
Timber battens
Services space
13mm
plasterboard
ceiling
Composite hollow upper floor with battered ceiling

10.4 Plank floors

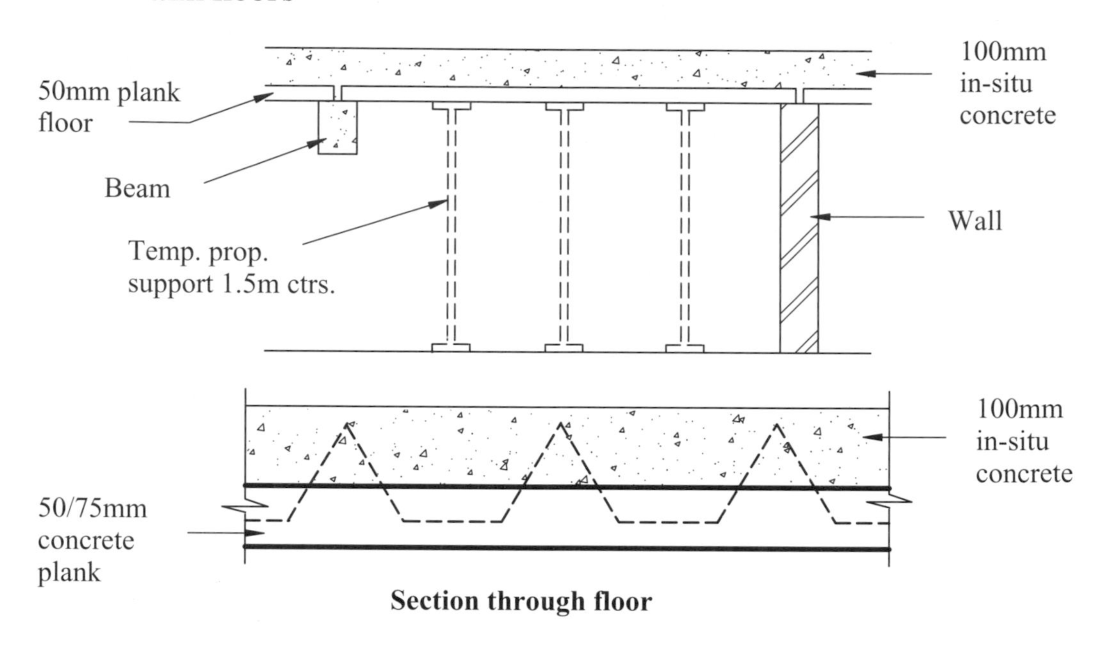

Section through floor

Precast concrete planks waiting to be positioned at the next floor level once temporary support erected.

The 75mm plank floor shown in position. Plank floor acts as permanent formwork to support the in-situ concrete slab which is poured over. The 75mm-thick planks do not require any propping.

Precast beams are being positioned in a basement area to form the support for plank floors. Commencement of temporary propping to the floors is shown adjacent to the wall in the centre of the picture.

Plank floors are extensively used in the Netherlands, Denmark and Germany in crosswall construction.

CHAPTER ELEVEN

CLADDING BUILDINGS

11.0 Overview

A range of case studies are illustrated in relation to the methods of cladding a range of medium and high-rise buildings.

Case study 1–Spinneyfield project

A twelve-storey modern office block of storey-height cladding panels handled by means of a mono-rail system.

References are included on the use of mast climbers for handling curtain walling and facing panels. The benefits of using mast climbers are illustrated.

Case sudy 2–four-storey office block project

The use of mast climbers is illustrated for the handling of both curtain walling and access to the fixing of panels. The example illustrates the use of canti-boxes in conjunction with steel stillages for the handling of the external storey-height panels.

Case study 3–four-storey college building

This case study involves the fixing of rainscreen cladding to structural insulated panels (SIPs). The lightweight zinc cladding is fixed from a scissor lift platform, obviating the cost of an expensive external scaffold.

Case study 4–12-storey retail and residential accommodation block

The iconic building on the waterfront in Liverpool is one of three similar finished blocks. A method statement is indicated for the handling of the storey-height panels, using a tracked hydraulic crane situated one floor above the panel handling level. This allows the reader an insight into modern handling techniques.

Case Sstudy 5–external brickwork cladding to an 11-storey frame

An interesting case study based on a restricted-access site. The eleven-storey precast crosswall frame is to be clad in faced brickwork panels. The external wall insulation and brickwork is to be fixed from mast climbers.

SIP Build–structural insulated panels

A section has been included on the application of SIP panels for use on commercial buildings, schools and supermarket projects. Information has been made available from McVeigh Insulations, fabricators and providers of large prefabricated SIP panels.

The use of structural insulated panels is likely to become an economic choice for single-skin construction in housing and multi-storey residential low-rise projects. Refer to McVeigh Insulations web site.

The SIP Build website allows access to a wide range of construction detail sheets for low-rise housing.

11.1 Cladding buildings

There are various methods of handling and fixing cladding panels to the facade of a building. Factors influencing the handling and fixing of the panels include:

- size and weight of the panel
- proposed method of fixing suggested by the subcontractor (subcontractor's method statement)
- plant proposed for handling the panels–i.e. tower crane / small mobile crane / monorail or mast climbers
- access to building face / boundary restrictions / proximity of other buildings
- whether an external scaffold is proposed or the panels have to be fixed from the internal floors of the building.

Fixing small panels

- Small individual panels may be fixed from external scaffold or alternatively from a mast climbing platform.
- On buildings up to 4/5 storeys in height, hydraulic platforms (scissor lift platforms and others) may be used.
- Brickwork panels, laid in situ, may be accommodated from a mast climber platform as shown in case study 5.
- Lightweight aluminium cladding panels fixed into the face of the building may be fixed from a scissor lift as shown in case study 3.

Fixing and handling storey-height panels

This applies to heavy cladding panels held in position by steel plates and connectors fitted to the ends of floors or to the wall face of a building.

The aluminium framing of a curtain wall system may be fixed from a mast climber platform. The glass storey-height panels may be lifted into position by tower crane with the fixing gang working off mast climber platform–refer to case study 4.

Storey-height panels, fitted to the edge of floors at the top and bottom of the units may be handled by mini-crane as illustrated in case study 4 (fixing of granite facing panels at Mann Island, Liverpool).

Extensive use is made of canilevered boxes that can be fixed between the floors of a building. These can be for loading materials for fixing in the building or for handling cladding panels. Refer to case studies 1 and 2.

Large panels may also be positioned on electric trolleys at ground floor level and moved into position within the building. Vertical movement internally is via a materials lift or hoist located inside the building. Refer to case study 4.

Specialist subcontractors are constantly devising alternative methods of handling materials in a more efficient way.

Cladding buildings in practice

Rainscreen panels fitted from mast climbers

Storey-height panels handled with a monorail system

Curtain walling fixed from mast climbers
Storey-height infill panels positioned by tower crane

11.2 Cladding case studies

Case study 1–Spinningfields project

The 12-storey inner-city building is clad in glass storey-height panels using a monorail system for handling the panels. Zena box platforms have been located at different floor levels on the building facade in order to facilitate the lifting of the panels.

Six floors of the building facade are shown with every second floor clearly defined with a dark glass strip.

Projecting deck box platforms are shown extending from the facade of the building. The monorail can be seen lifting a panel from the Zena box.

Stages involved in handling the glass storey-height panels

- Delivery of the panels to site and transfer to electric trolleys at ground-floor level. The trolleys are then moved into a central building hoist for raising to the appropriate floor level.
- At each floor level the trolleys are moved onto the Zena platforms for lifting into position with the monorail equipment.

On a similar project involving the fixing of storey-height panels a small hydraulic crane lifter was used instead of the monorail to lift and position the glass panels from the projecting platforms. The small crane could readily be moved from floor to floor as work proceeded at different storey heights.

Storey-height panels are shown in position on other floors of the building.
Refer to case study 4 where similar trolley systems were used on the Mann Island project.

Using mast climbers for cladding buildings

General view of storey height panels in position on a lower floor

A range of situations are illustrated involving the use of mast climbers for fixing of both small and large cladding systems to facade of both medium and large high-rise buildings.

A single mast climber being used on a ten-storey building for fixing the light metal cladding panels is illustrated, as is a double mast climber with an extended corner return arrangement in place.

Mast climbers are mainly used for fixing of lightweight fixing panels with access for up to four fixers. Heavy panels will have to be handled by crane or special lifting equipment.

Large external timber panels are being raised on a lift-climbing platform prior to fixing them vertically between floors. Also shown are two mast climbers fixed to the front face of a four-storey building. Note that certain small panels have been left out to accommodate the fixing of mast climber ties to the building frame.

Single mast climbers

Double and single mast climbers with return end

Benefits of using mast climbers

SIP panels being positioned from mast climber platform

Mast climber tower tied into face of building at each floor level

The benefits of using mast climbers

- Reduced access time
- Number of building ties used
- Improves safety during and after working hours
- Extension platforms available can contour most buildings
- Free-standing mast climbers up to 30 metres high
- Less damage to components when handling and fixing
- Reduction in the fixing period by up to 40 per cent
- Weather screening improves productivity

Time and cost savings

- Materials and operatives can lifted at the same time
- Project time reduced when compared with normal scaffolding
- Quick and easy erection and dismantling process

Technical advantages

- Less restriction on height–up to 30 metres maximum
- Platform lengths may vary from 2 to 20 metres
- Self-propelled wheel trolly arrangements allow mobility around site
- Variable-width platforms available
- High payload for single mast climbers
  i.e. 5000kg and 10,000kg for double mast climbers

Case study 2–four-storey office block

Fixing of curtain wall supports from platform

The project involves the use of double mast climbers for the fixing of the storey-height glass panels.

The rear elevation of the building is shown with two mast climbers in position.

The glass cladding panels are delivered to site on metal-framed glass stillage platforms. The stillages are lifted directly onto cantilevered deck platforms as shown and moved into the building for storage. The platform stillages were also stored on the roof of the building for ease of handling with the tower crane.The site space available was extremely limited as no external storage area around the building perimeter was available.

The aluminium framing is positioned ready to receive the storey-height glass panels. The glazing frames have been fixed from mast climbers platforms.

The steel stillage is shown in position on one of the cantilever decks. The storey-height glass panels are lifted into position from the platforms by tower crane. This type of Canti-deck platform is extremely popular in the construction industry.

Canti boxes for materials handling in position on building facade

Curtain waling support in position

Steel stillages for handling external glass panels

Case study 3–Stockport College project

SIP's fixed to steel frame.

Aluminium support to lightweight panels

Scissor lift in use for fixing of panels

The project involves the construction of a four-storey-steel framed building clad in narrow zinc panels capable of being fixed from a scissors platform.

The panels are fixed to a story- height SIP (special insulated panel) which is fixed to the external steelwork. The insulated panel consists of two skins of 20mm thick stirling board with a blown foam insulation core. The large SIPs have been lifted into position using a 360° slewing forklift truck arrangement.

External SIP's are shown in position along one facade of the building. Aluminium fixing support rails have been positioned to receive the vertical zinc panels.

Fixing of the zinc panels was undertaken from a scissor lift.

Vertical zinc panels were fixed between the external windows. The choice of cladding material has a distinct impact on the fixing costs as it can be easily handled without the use of complex handling equipment. SIPs allow lightweight rail screen protection systems to be incorporated into the design.

Vertical external panels in position

Case study 4–12-storey residential block Liverpool

Architectural impression of completed building

Granite-faced aluminium framed panels

The 20mm granite facing slabs were mined, cut and polished in China and shipped to Italy where the insulated aluminium backing panel was fixed. They were then crated and despatched to Liverpool.

On site the plastic wrapped panels are lifted from the crates onto powered trolleys.

The panels are then moved by goods hoist to the appropriate floor level where they are to be fixed.

A method statement extract is included to indicate safety hazards to be addressed during the handling and erection of the panels.

Slot-in brackets are fixed to end of the in-situ concrete floors. They allow each panel to be held in position at each corner.

External windows to each apartment are formed within the panels.

These appear black externally but allow the occupant to look out through an opaque window.

Base fixing bracket

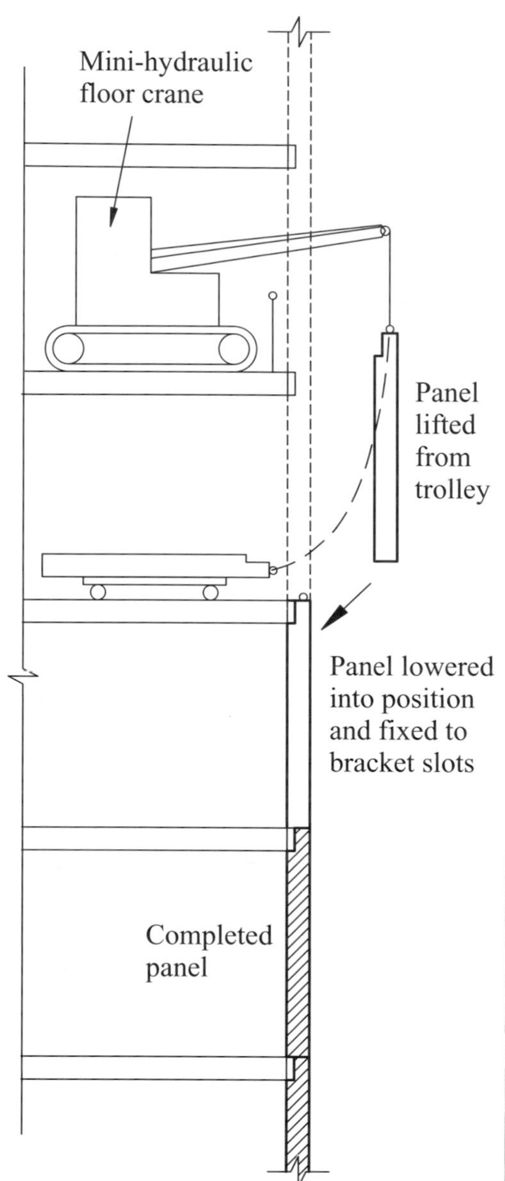

See separate method statement notes from installer

Fixing brackets at each of the corners of the panel

Bracket tied back to floor slab

Method statements extracts

The method statement extracts are from a typical work package subcontractor supplying and fixing the granite panels.
It has been somewhat shortened to highlight the main areas of the handling and fixing process.

The method statement should address the following areas relative to the operation being undertaken.

- Description of proposed methods to be adopted for panel installation.
- Highlight specific safety measures to be adopted to ensure safe working conditions for operatives and plant.
- Sketches may be incorporated as appropriate to the task.

Panel installation

For panel installation, access to the brackets at the bottom of the panel and access to the supporting brackets at top of the panel is required therefore the following procedure will be carried out:

- A safety line will be spanned between columns at high level.
  Certified slings and lines will be used. The safety line may already be in place as previous line and level of the brackets has been carried out.
- Access to the fixing area will be restricted by using KGuard leading edge footplates and standard guard barriers. These will also provide a toe board.
  The footplates and posts are prepared for fixing the guard barriers at the relevant area.
  This is only a secondary safety measure as access to the floor will be restricted.
  Only fixers and other operatives with permission will have access to the floors.
  The floor where the edge protection has been altered will therefore be cordoned off.

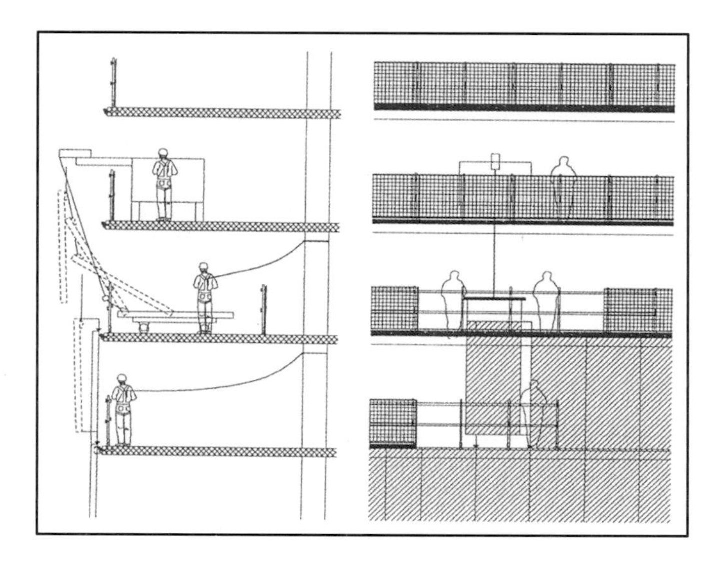

Sketch : Setting out of KGuard barrier and link bars for panel. Installation showing also location of floor crane and panels.

- The fixers will be harnessed and connected to the safety line by the use of adjustable rope graps. The length of the rope will be adjusted when the edge protection is still in place. The safety line works as a fall restraint system and will ensure that the operatives cannot fall over the edge of the slab. Prior to this system being used, each operative of the two involved will ensure his workmate is attached to the adjustable restraint line prior to the removal of any edge protection.

- The guard barrier will then be removed and used at the prelim posts already prepared to provide an exclusion zone.
- Prelim link bars will be fixed to the posts at the slab edge. This provides proper edge protection again and the safety line is now only working as a secondary measure. This preparation for getting access to the slab edge will be carried out on the actual floor, where the panel will be installed and at the floor above as fixers will guide the panel into its final position from behind the edge protection at both floors.
- Access is now possible through the link bars.
  Either the operatives will reach through between the link bars kneeling on the floor slab or will get access below the lower link bar, lying on the floor.
- Two floors above the designated panel position a floor crane will be positioned. Two possible crane types may be used on this project:
  –The Valla 20 E Mobile Crane
  –PK Kupfer Mobile Crane, customised solution

  Loadings have been approved by the main contractor prior to the commencement of work.

  Both cranes need to be hoisted by the goods hoist and need a 110 V power supply.

  Floor crane operator will be trained prior to starting the installation process.
- With the floor crane in position (min. one floor above the actual panel location) and the lifting arm reaching over the KGuard, the actual installation process may start. The crane hook and lifting bar are connected to the T-brackets at the top of the panel.
- At the floor, where the panel is prepared in a horizontal position on trolleys, the link bars off the edge protection will be locally removed either both or just the top link bar. This is necessary to move the panel out of the building (see sketch). The panel will be transferred to the outside of the building by lifting the top of the panel and guiding the bottom of the panel on the trolleys until the panel is hanging in vertical position.
- If necessary, the panel will be turned by 180° so that it is correctly orientated, ready to be connected to the structure.
- Finally the panel will be moved to the designated position, all in accordance with the approved drawings, and will be connected to the prefixed brackets already connected to the structure as described in the bracket installation method statement. Operatives will guide the panel into position at top and bottom.
- When the panel is located in its final position and connected to the structure the lifting beam will be realised.

Case study 5–Eleven-storey hotel project

Brickwork cladding to a precast concrete frame

General view of eleven-storey precast concrete crosswall frame in central Manchester.

Site has very restricted access from all sides, one side bounded by a river with a limited access roadway on the opposite side.

The main building frame is of precast concrete crosswall construction. Details of the typical wall and floor connections are illustrated in chapter 7.

The external brick cladding and external windows are being fixed from mast climbers working on each facade of the building. The precast crosswall frame was erected at a build rate of one floor per two weeks.

The choice of external brick facings for the cladding of the frame is perhaps the right choice of finish for inner-city Manchester.

As an alternative, the facade may have been finished in a rain screen protection system fitted on an aluminium framework to the face of the building.

Brickwork cladding to PC frame

200mm-thick precast concrete wall to be clad externally in dark-coloured facing bricks

Position of spacers to receive 70mm rigid insulation slabs

Aluminium angle support fixed to face of precast wall–brickwork to be coursed between lower support and angle support to next storey

Use of mast climbers for access to fixing insulation and brick facing to wall panels. All materials and labour accessed by mast climber platforms.

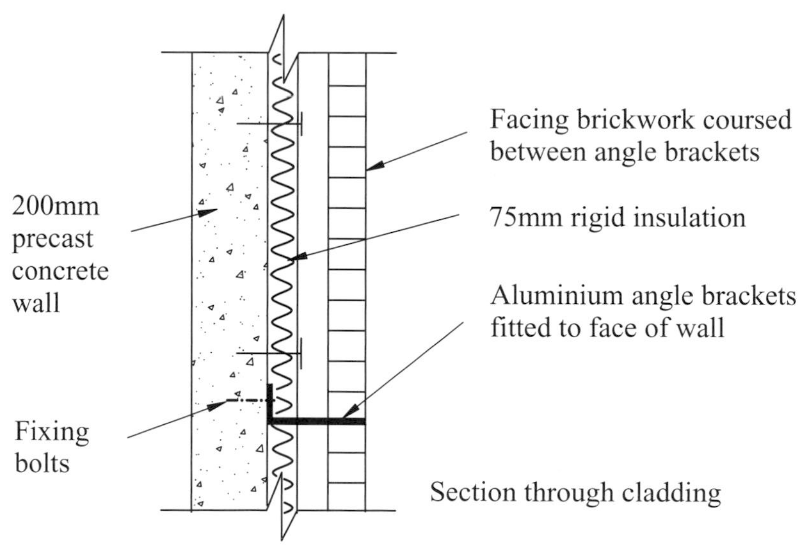

Section through cladding

11.3 Structural insulated panels

The use of SIPs is becoming increasingly popular for a wide range of structural wall applications. They are an innovative method of construction allowing the rapid erection of buildings for both domestic and commercial use.

SIPs provides an ideal backing for overfixing rainscreen cladding directly to the face of the panels. In certain cases they may eliminate the need for structural frames. They have been used extensively in construction with glulam frames. Offsight production of large prefabricated SIPs incorporating window and door openings has speeded up the construction build time. The use of SIPs is becoming increasingly popular for time-sensitive projects.

A range of case study photographs are included for projects undertaken by McVeigh Insulations.

The key benefits of using SIP panels on various projects are summerised. Based on extracts from case studies presented on the McVeigh Insulations website.

Structural insulated panels (SIP)

Application to multi-storey building

Use of SIPs to receive rain screen protection cladding
to multi-storey concrete frame.

Extract from McVeigh Insulation Ltd
website: www.mcveighinsulation.co.uk

What are SIPs?

SIPs are structural insulated panels used to construct buildings.
SIPs have been used extensively in the USA over the past 50 years but are now becoming a popular choice in the UK for timber-based construction. SIPs are environmentally friendly and economically sound. SIPs are the innovative building construction method of the 21st century allowing the rapid development of buildings for domestic and commercial use.

Schools, hotels and retail developments are realising the benefits that SIPs can bring to a project, and we are providing SIP solutions for buildings reducing or in some cases eliminating the need for a structural frame using the inherent strengths that a SIP can provide.

A SIP provides an ideal substratum for overfixing rainscreens. This can be done on site in the traditional manner or off site providing quicker system build techniques. These modern methods of construction (MMC) solutions are becoming increasingly popular on time-sensitive projects.

How are the SIPs made?

A basic SIP panel is made from orientated strand board (OSB) facing boards with CFC free/ODP zero closed-cell polyurethane core. SIPs can be used to constuct the floor, walls and roof of a building, enabling detailing at interfaces providing continuity of insulation and minimal air leakage.

Historically, sandwich panels have been used in the UK mainly for buildings with high thermal requirements. As the Building Regulations are demanding more efficient structures the sandwich panel solution is becoming more favourable.

Why choose SIPs?

SIPs are used all over the world because of their superior strength and quality build. Homes and commercial buildings that use this innovative technology have survived hurricanes, earthquakes and tornadoes, when other buildings around have not. In the UK, SIPs are more capable of withstanding climate extremes as well as maintaining their strength throughout the years and providing a safe and secure living environment.

Benefits of SIPs

- Environmentally friendly

  SIP timbers are used from sustainable sources and are made with ODP-zero insulated core.

- Speed

  Quick installation is made possible by using large prefabricated units. Envelope completion allows following trades to start earlier.

- Strong and lightweight

  Rationalise foundation design. The nature of the panels means that open spaces can be constructed without the need for additional framing.

- Rooms within roofs

  Fully utilise the roof space of buildings with the long span capabilities of SIPs.

- Defects and shrinkage

  Being made in factory conditions helps to reduce defects. Comparable to timber frame construction, shrinkage and subsequent making good are minimal.

- Minimal waste and maximum recyclability

  All panels being cut to length and apertures formed in the factory results in minimal site waste. Off-cuts from engineering panels in our workshop can be recycled.

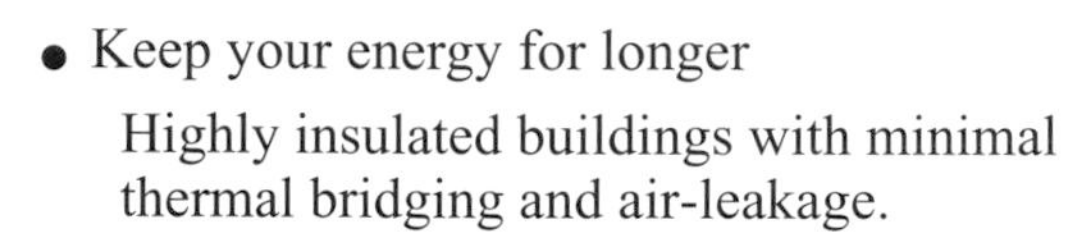

- Keep your energy for longer

  Highly insulated buildings with minimal thermal bridging and air-leakage.

- External finishes

  Many external finishes can be applied to SIPs, from render, rainscreen and weatherboards to brick slips, tiles or conventional brick outer skin. All can be applied off the critical path of the overall project duration.

11.4 SIP Panels–McVeigh Insulations McVeigh SIP fabrication process

The design team

3m long by 1200mm wide panels being assembled. Panels to be used for external enclosure of a six storey steel frame. Panel surface protected with a breather barrier and stored on metal racks ready for delivery to site.

SIP panels as purchased in specific lengths from manufacture

Forming openings in prefabricated panels.

Forming rebates to the panel edges

Fabrication and inserting of spines and strengthening posts and openings

Handling and fixing SIP panels

Panels lifted from ground level with spider crane located inside the building at appropriate floor level

Panels stored on vertical racks. Moved into position at ground-floor level with hydraulic lifter

Panel fixers working on face of panels from hydraulic platform

SIP panels in final position– fixed at bottom and top of units to steel frame

SIP panels to six-storey college building.
Panels cladded externally with zinc rainscreen protection lightweight panels.

Panels positioned on top floor of building, lifted with spider crane from ground level

The SIP subcontractor was invited by the Architect to assist in assessing the value engineering aspects of the external cladding.

SIP were selected to replace the lightweight steel framing infill. In addition to significant cost savings additional benefits include:

- higher degree of control with regard to quality and consistancy
- off-site fabrication reduces the overall programme and working at heights
- panels provide high thermal efficiency and low air leakage
- minimum movement due to composite nature of panel construction
- rainscreen cladding can be easily fixed to the surface of the panels. The cladding may be fixed from a scissor lift or hydraulic platform or by using mast climbers. This results in further savings in external scaffolding costs
- windows and door openings are framed out with structural timbers, eliminating the need for secondary steelwork

Part completed elevations of building

CHAPTER TWELVE

TIMBER-FRAMED CONSTRUCTION

12.0 Overview

The various forms of timber-frame construction used in the UK (as specified by TRADA) are first outlined. These mainly relate to the application of the platform frame method of construction. Reference is also made to the suppliers of post and beam frame types. The advantages and disadvantages of timber-framed construction are summarised.

Images are shown of the site build platform method and large panel platform frame (case study in France).

Construction detail is provided for timber-framed construction together with details from author's own timber framed-home in Derbyshire.

The use of structural insulated panels (SIPs) for housing construction is also illustrated both from a construction detailed stage and in relation to a SIP Build project in Cyprus. The application of finishes to external and internal walls is outlined together with the advantages of SIP construction for housing projects.

12.1 Timber-framed construction

Notes summarised from TRADA Frame Construction 2nd Edition 1994

The platform frame

The platform frame is the most commonly used method of timber framed construction in the UK. Each storey is framed up with the floor deck of one floor becoming the erection platform for the next.

These approaches are outlined by TRADA as:

Platform frame

- Site build platform frame where the timer panels may be constructed on site.
- Small panel platform frame–factory-made panels up to 3.6m in length designed to be manhandled into place or by forklift.
- Large panel platform frame–factory-made full-width panels designed to be lifted into position by crane. The panels may incorporate ancillary components, e.g. fitted window and door frame.

The figure on page 313 provides simplistic diagrams of each type of platform frame.

Other methods of timber-framed construction include post and beam frames and volumetric housing systems.

As timber-framed construction has been developed, platform systems incorporating post and beam principles have been introduced. This enables the centre of the building to be opened up to provide an added design feature. (refer to example on author's own house) as illustrated.

Post and beam frame

The frame has been developed from the Elizabethan period oak-framed modular house. Many barn type buildings and cottages have been constructed of large sectional timbers and infilled with brick or wattle and daub type materials in the past.

The modular post and beam frame utilises engineered glulam beams (laminated timber beams) to form a structural skeleton. This is then filled in with non-structural engineered panels. Structural insulated panels (SIPs) have recently been developed which meet current building regulations.

The principles of the post and beam system are illustrated.

Alternative forms of the panel system

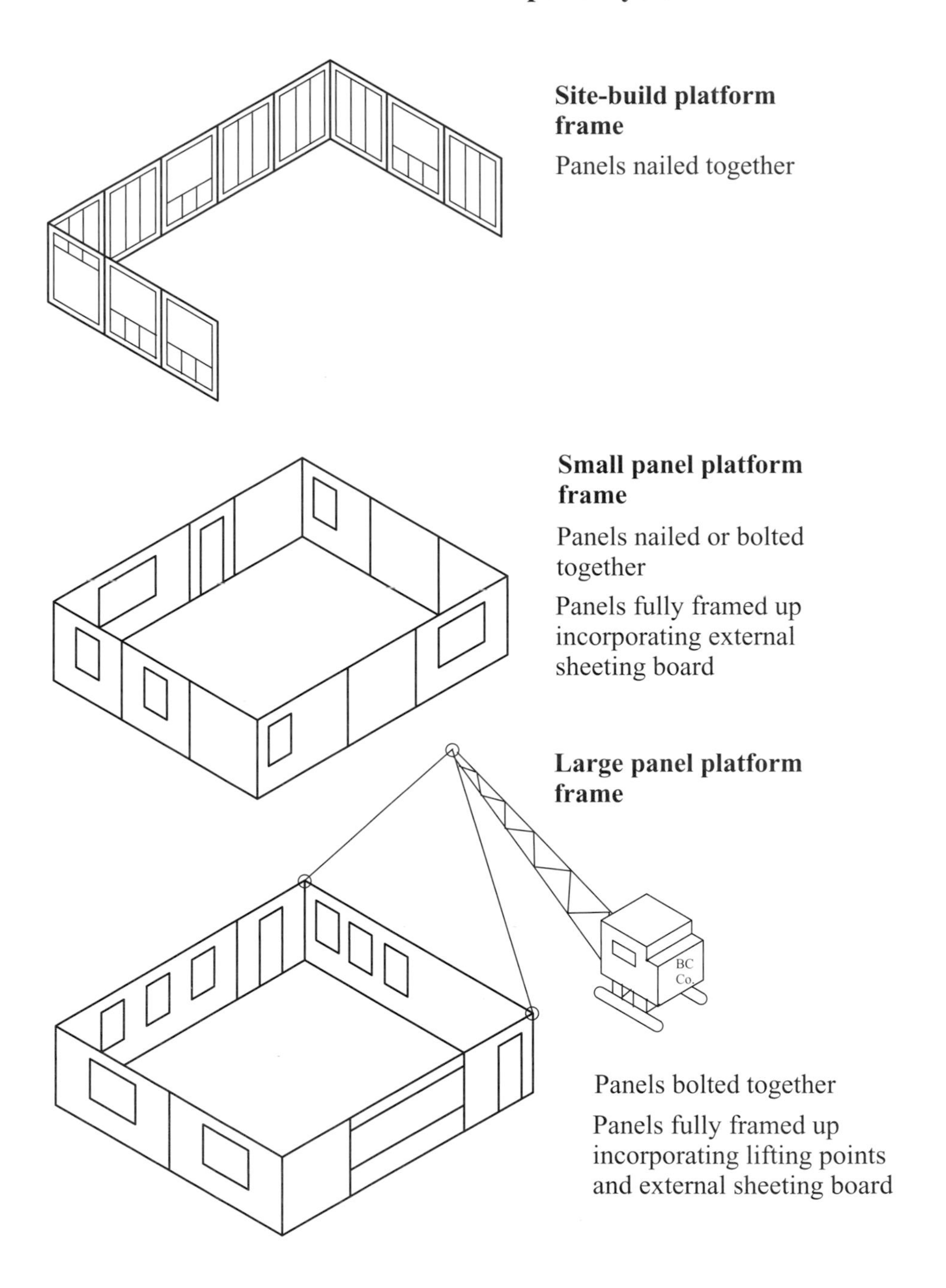

The post and beam frame is offered by many UK timber frame companies including:

Polton Limited
Benfield ATT Modern Post & Beam
Christian Torsten Timber Frame
Davies Frame Company
Welsh Oak Frames
Border Oak
Goodwins Timber Frames

The Huf Haus post and beam frame incorporates the use of high specification laminated construction, factory produced in Germany. An erection period of 10 days is achievable for the complete structural frame incorporating small panels, fully glazed wall panels, roof construction and tiled roof.

Advantages of timber-framed construction

Manufacturers' literature, catalogues and specialist self-build publications outline the pros and cons of timber frame construction. Here are a range of extracts from such material.

Advantages to

The designer	–Design flexibility –Variety of forms of construction, and mix and match approach with traditional finishes –Use of engineered structural components e.g. glulam, laminated beams to increase spans –U values significantly improved –Reduced foundation costs due to lighter building loads
The developer/ builder	–Rapid erection sequence for substructure and roof –Timber frame two-storey units–erection time 10 days including roof. Traditional construction 50–70 days
Improved cash flow for developer	–Shorter contract period results in fewer preliminaries and lower supervisory costs which may not directly benefit the client/owner –Less wastage due to the use of modular components–less cutting waste etc. –Simplicity of construction process resulting in savings in labour and plant costs

To the owner –Individual client requirements may not cost the earth
–Lower heating costs due to the extensive use of modern insulation materials
–Wide choice of specification levels resulting in clients cost savings

Disadvantages of timber-framed construction

On paper there does not appear to be any! This is not so, having built a timber-framed house and lived in it for 10 years.

Typical experiences of living in a timber-framed house include the following niggly problems.

- Fixing hooks/pictures/furniture etc. to partition walls. Solution: have additional horizontal timber studs inserted at worktop and picture rail heights to ease the fixing support problems. Record the height of these rails to aid locating them.
- Sound insulation between rooms and floors can be a problem–one has to learn to live with it.
- The house tends to overheat too easily, therefore install a heating system which is easy to control. Provide plenty of through-room ventilation.
- Incorporate Velux roof lights in sloping roof areas, to efficiently cool the place down and provide through passage of air.
- Don't incorporate too large a wood burning stove - if you do, you will have to sit at the back of the room once you light it (i.e. 3–5kW output only).

Timber frame statistics

In 2006, the total number of new houses constructed in the UK was 215,000. Twenty per cent (approx. 40,000) were constructed using timber frames. The percentage use in Scotland, however, peaked at 80 per cent of the houses constructed. The self-build market is predominately of timber-framed construction and the introduction of SIP-built houses will influence the market in the future. The self builder has extensive source information available in the form of monthly publications such as *Build It* and *House Renovations.*

Post and beam frame

(Similar to a concrete column and beam frame–Timber sections 200 × 200mm columns)

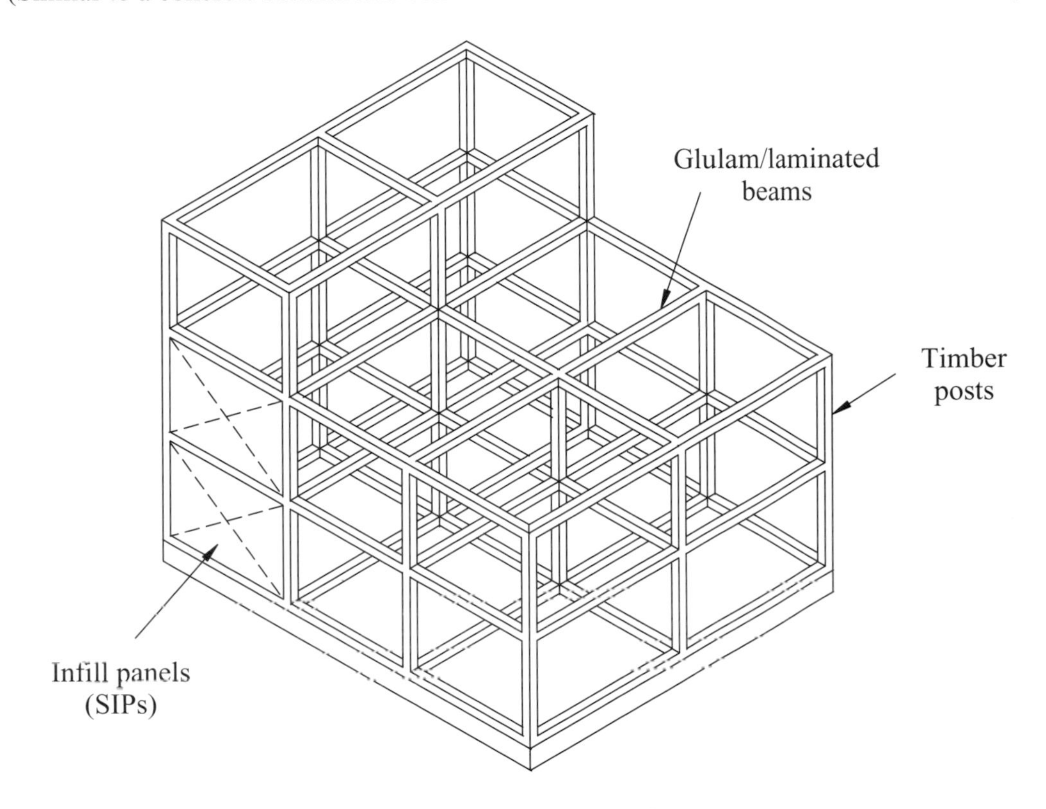

Modern post and beam frame

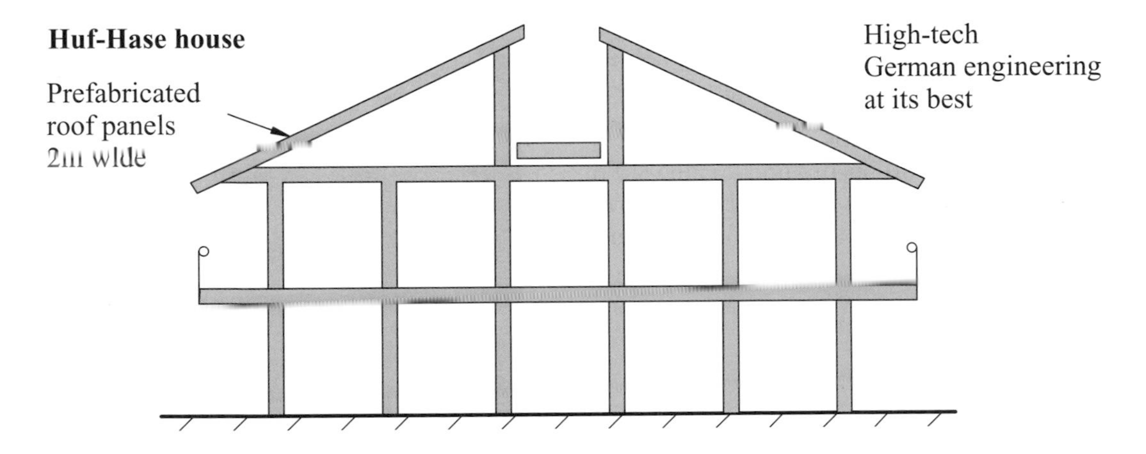

12.2 Site-build platform frame

Front Elevation

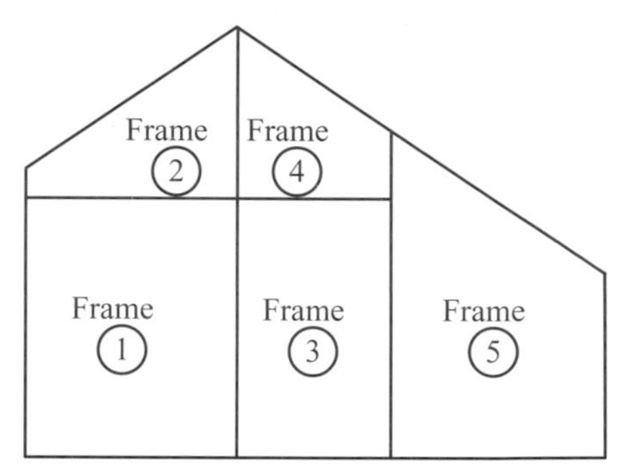

Front elevation
Panel arrangement
Fixing order of panels

Front and part side elevation

Erection of a single-storey garage
to a domestic house

Panel 5
All panels constructed of 50mm
x 125mm sections

Elevation of front of timber- framed house

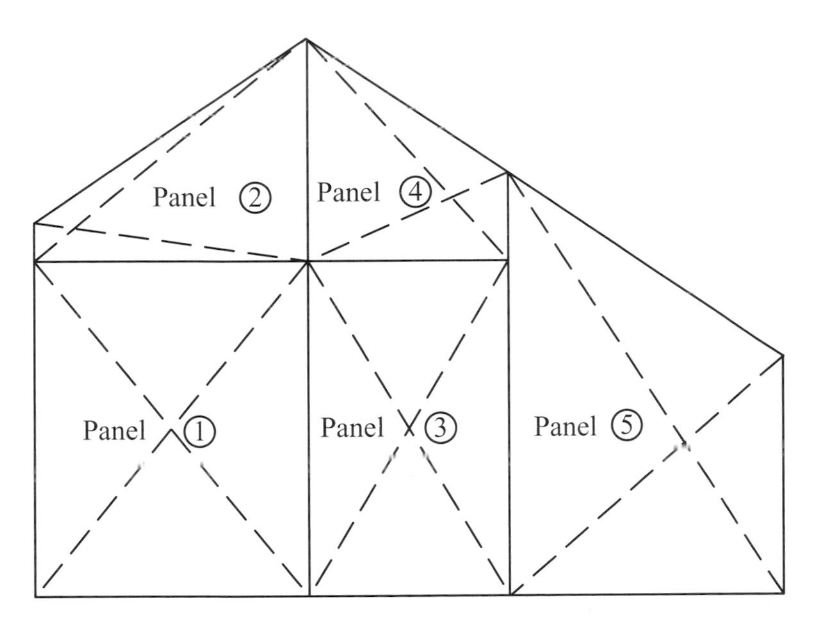

Sequence of fabrication and erecting panels

Project: construction of a garage structure–domestic construction.

The timber-framed garage structure 7.2–5m on plan. An elevated storage platform is located above the garage door entrance area.

The front elevation is built up of five panels prefabricated on site. The rear elevation is of similar layout.

The single panel construction is built up as shown.

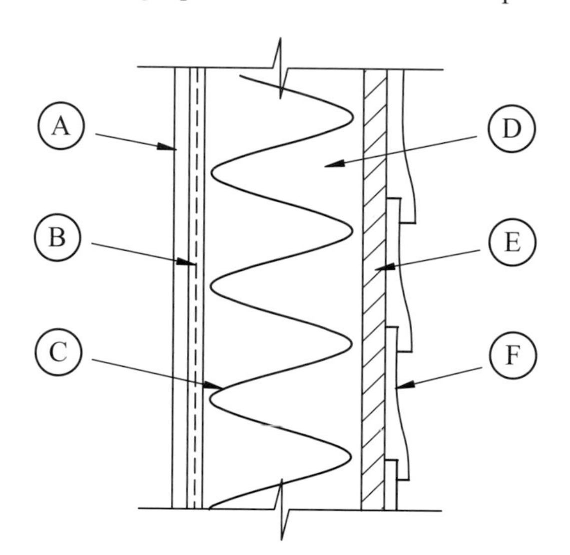

A –13mm plasterboard internally
B –Vapour barrier
C –100mm insulation
D –50 x 125mm timber studs
E –20mm stirling board
F –Cedar shiplap boarding externally

Exterior wall section (non-habitable room)

The roof is to be finished in clay tiles to match the adjacent dwelling.

Overall construction period–self build

Foundation flat slab	3 days
Fabricate/erect frame	10 days
Wall cladding	4 days
Roof cladding	4 days
	21 days (total)

12.3 Small panel platform frame

Two-storey house–sequence of work

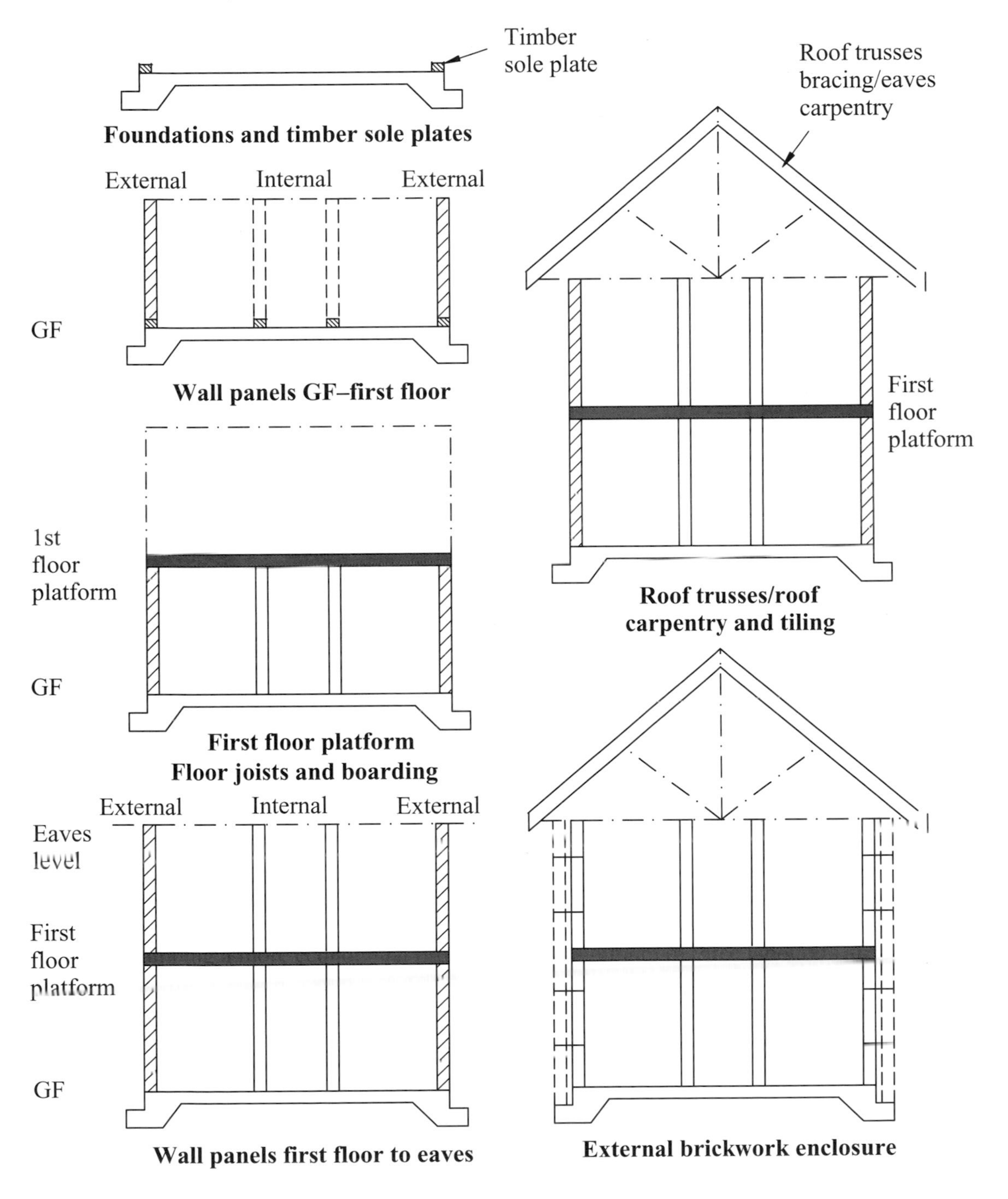

Foundation–beam and block floor

External scaffold in position from ground level to eaves

Wall panels GF to first floor, followed by first floor platform

First floor to eaves wall panels and roof trusses in position

12.4 Timber frame construction detail

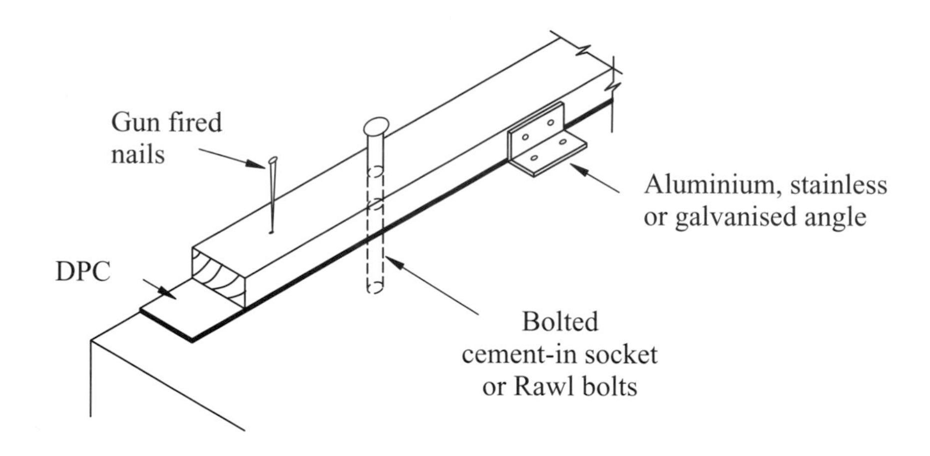

Fixing of sole plate to foundations

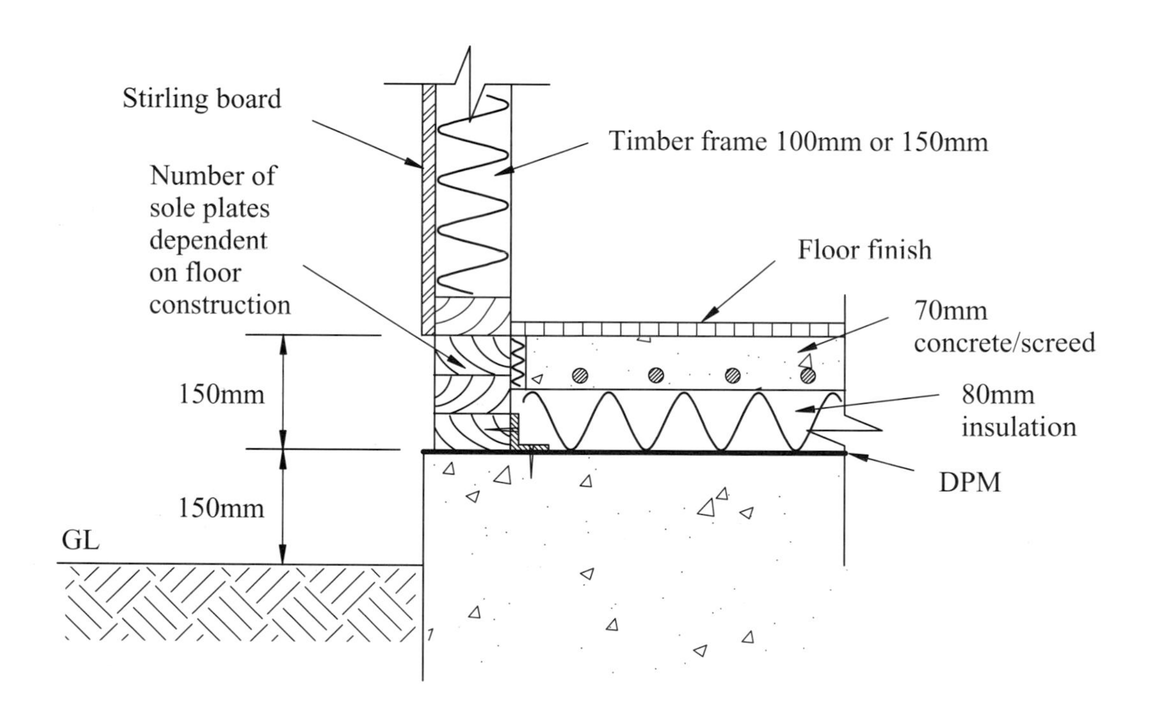

Section through floor incorporating underground heating

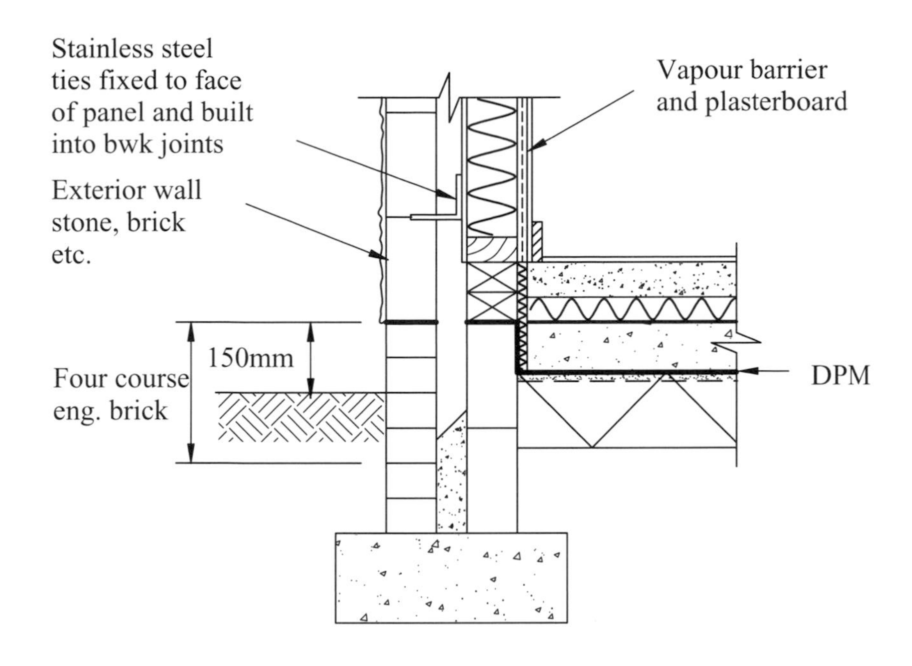

Foundation section

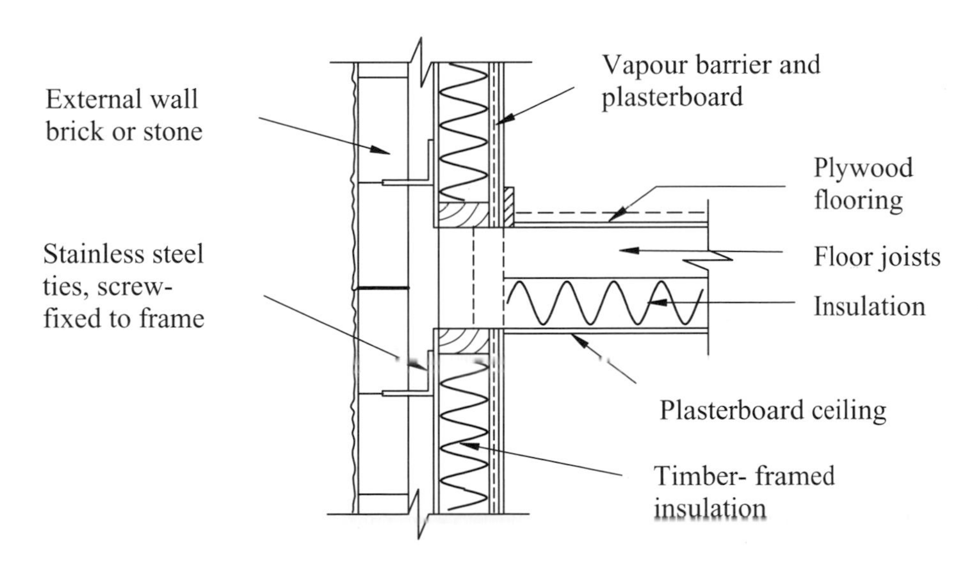

First floor section

Timber frame construction detail

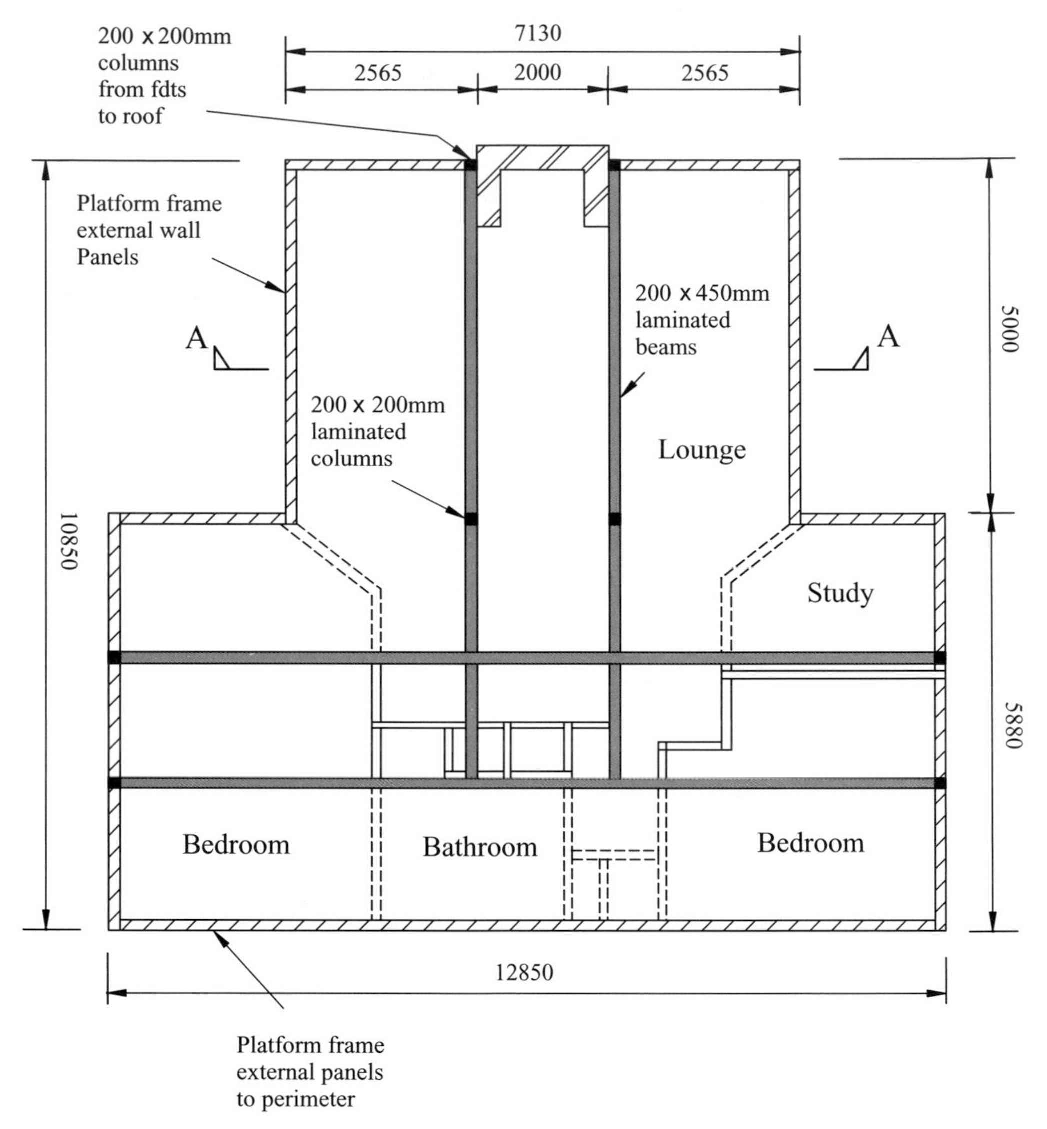

First floor plan (Author's home)

Layout of laminated beams and columns showing support to roof structure at first floor level. Platform frame used for ground floor to first floor construction.

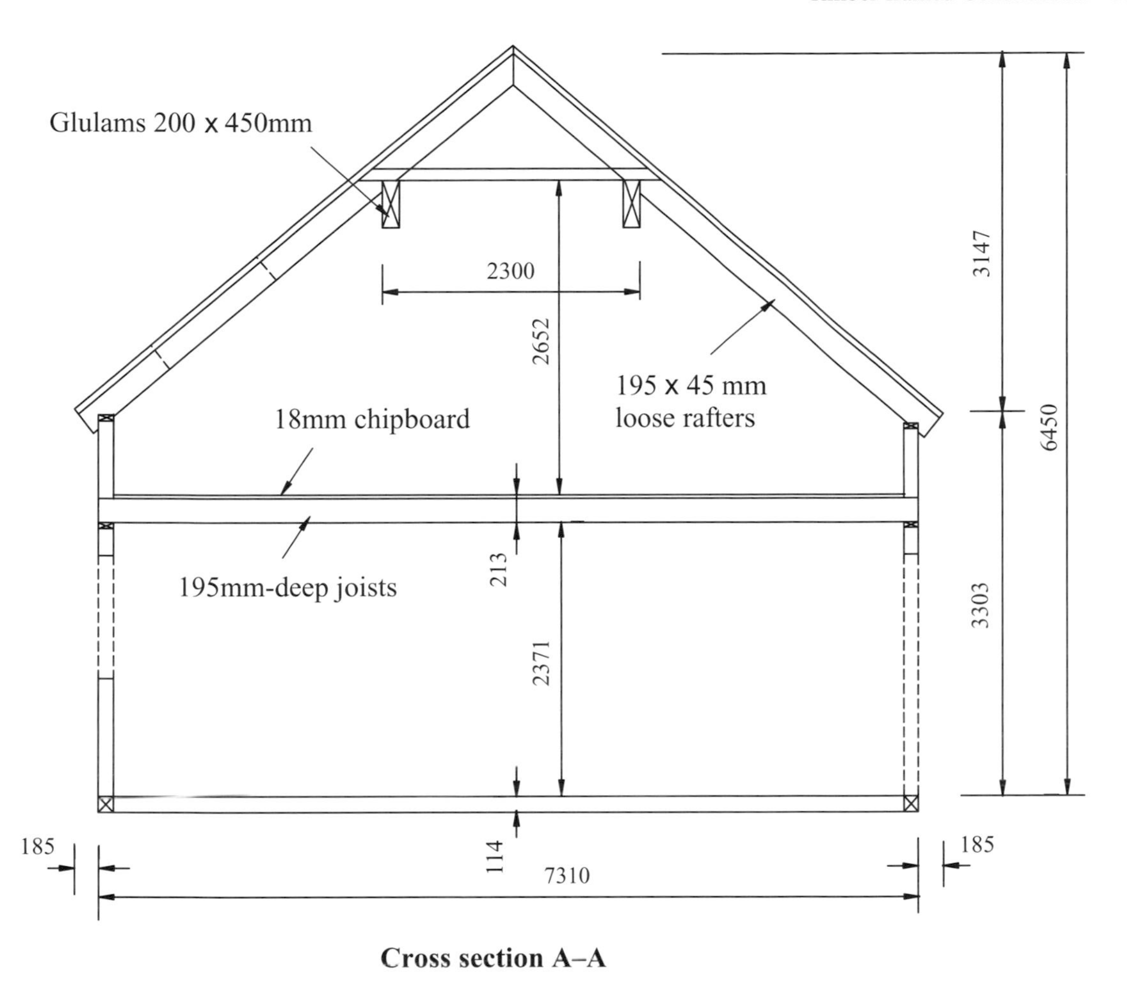

Cross section A–A

The timber-framed house was constructed in an overall period of 9 months.

Foundations	6 weeks
Frame and roof	4 weeks
External stonework	6 weeks
Internal finishings	12 weeks
Landscape	6 weeks
	34 weeks

The timber frame was supplied and erected by Taylor-Lane of Hereford. The company allow the purchaser to undertake any of the work packages involved in the construction. Refer to website.

12.5 Large panel platform frame

Erection of a timber framed house in mid France

Large wall panels to external elevations. Units craned into position with site-based tower crane.

Post and beam frame constructed internally to support first-floor platform.

Open-plan layout on ground floor achieved by opening up the central area of the house.

Large-panel timber-framed house constructed on a concrete raft foundation. Frame sole plate bolted to foundation slab. Panels bolted together as shown in details.

Internal view of post and beam frame structure supporting floors

External wall panel

Sole plate bolted to concrete upstand with DPM in position

Panels bolted together and galvanised bracket on frame connections

Bolted and galvanised angle connection

Connection between ends of floor joists to main beams

The continental timber-framed systems utilise much larger timber sections than those in the UK. Extensive use is made of bolted connections in securing the frames together and when fixing the frame to the foundations.

Internal view of post and beam frame to support first floor and flat roof

Platform frame for external walls and some internal areas. Post and beam frame constructed internally of laminated columns and floor beams. Internal post and beam frame shown

Internal view showing open space created to ground floor area

Post and beam frame to two-storey house

12.6 Structural insulated panels for housing

Background information:

SIPs were introduced into the UK in the 1970s. They have been established in North America and Scandinavian countries since the 1950s. Systems in the UK have adopted the construction and site practices of the European Union.

Description of SIP panel construction

The system mainly consists of a single skin panel which replaces the traditional cavity wall used for domestic construction in the UK.

SIP panels consist of two facings of 11mm OSB (oriented strand board) bonded by pressure injection to a polyurethane foam core. The overall panel thickness may be 100, 125 or 180mm. The facing and core act as a composite structure. All the timber in the main process is from sustainable timber sources.

Panels are up to 1200mm wide and 3000mm in height. Roof panels are available up to a maximum length of 6500mm. A single panel thickness of 100mm replaces the ceiling rafters on a traditional house and is structurally stable as well as providing an adequate U value to satisfy building regulations. Details of wall and pitched roof details are illustrated.

SIP build–construction detail

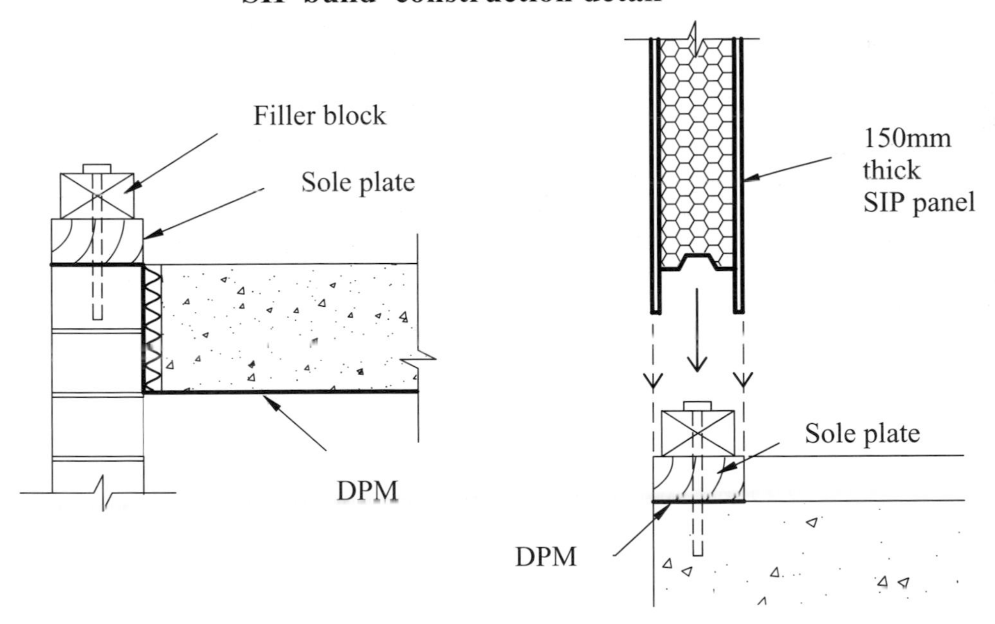

Fixing a SIP to the building sole plate

Application of SIPls to low-rise housing

SIP build–construction detail

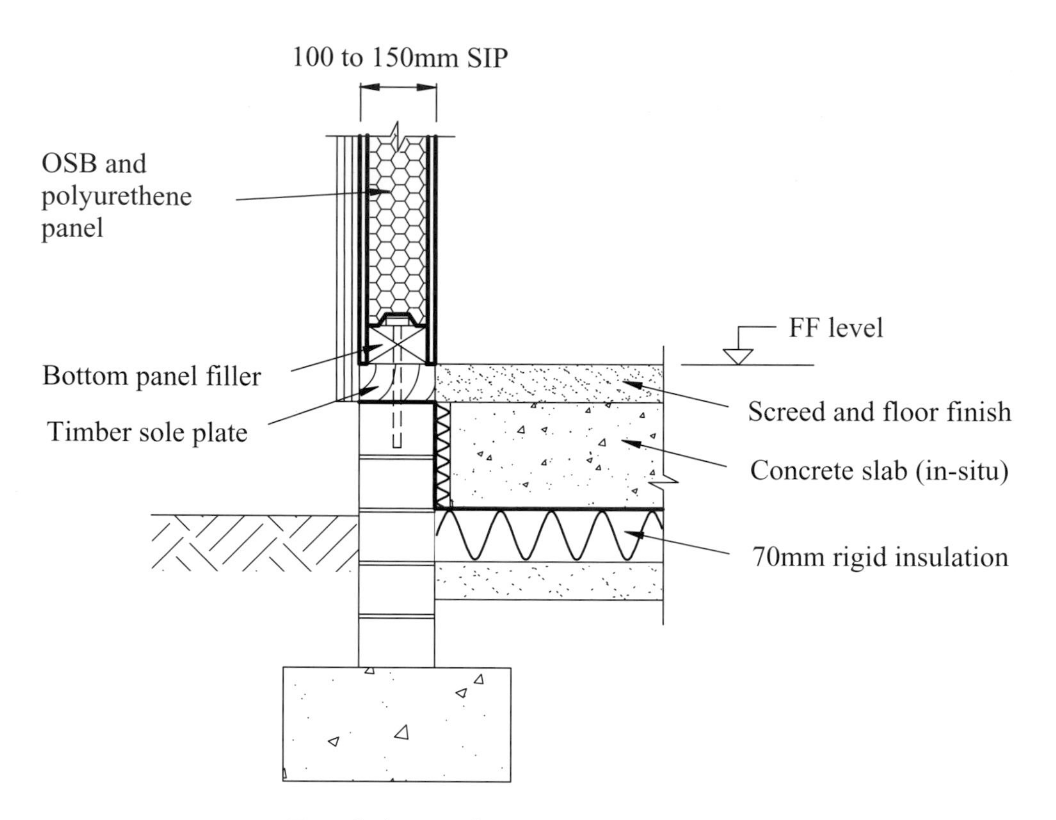

Foundation section

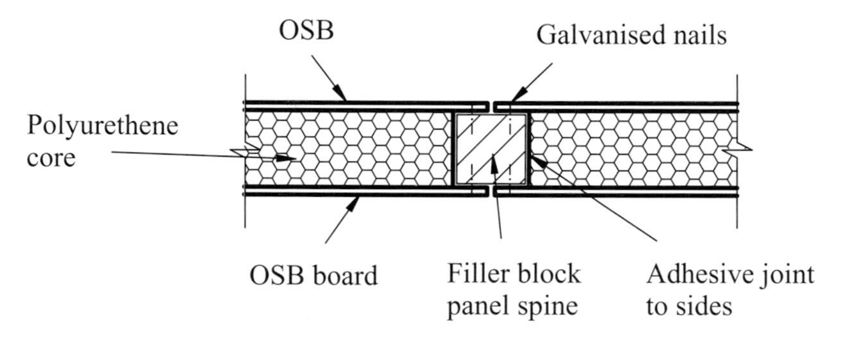

Connection detail between adjacent vertical panels

Application of SIP panels to low-rise housing

SIP build–external and internal wall finish

External wall finishes include:

- three-coat plaster finish on stainless steel lathing fixed to vertical timber battens
- external brickwork and cavity–stainless steel fixing ties nailed to external surface of panel
- sprayed textured coating applied to cement particle board fixed on 15 x 25mm vertical timber battens
- horizontal UPVC cladding fixed on 25 x 38mm timber battens (drained at bottom)
- brick slip system fixed on 25 x 38mm timber battens.

Internal wall finishes include:

- 12.5mm foil-backed plasterboard on 50 x 25mm battens (services may be incorporated behind plasterboard)
- 12.5mm foil-backed plasterboard fixed direct to panel
- double layer of plasterboard fixed direct to panel–rear plasterboard may be cut to form service chase.

On the SIP Board website www.sipboardltd.com construction detail is available covering all aspects of the building panel joints, floor and roof detail and finishes–an excellent website.

SIP build–advantages

- **Environmentally friendly**
  ODP-zero insulation core and faces made from sustainable forest resources.
- **Excellent thermal performance**
  High U value, lack of thermal bridging and airtightness.
- **Saving on construction time and cost**
  The use of large engineering components results in a shorter construction period. No heavy plant required for the erection process. Saving in time results in savings in labour costs and contract preliminaries.
- **Space maximising design flexibility**
  Open plan approach to internal layout. Sloping roof space allows for additional rooms in roof area
- **Lightweight structural system**
  Light foundation loading leads to saving in foundation costs and the simplicity of the foundations. Suitable for constructing on a simple raft slab.
- **Wide variety of external elevational treatments**
  Easy to apply finishes both internally and externally. Traditional brick and stone feature walls may be readily accommodated in the design. May be used as the internal wall in cavity wall construction.
- **Reduced heating costs**
  Reductions in heating costs due to high U value panels. Conventional heating systems can be used but they should be down-sized due to the house being so energy efficient, therefore reducing heating installation and running costs.

12.7 SIP build project–Cyprus

Private villa construction

The photographs were taken during the erection of a SIP build house/villa in Cyprus. They illustrate the simplicity of the SIP build system for low-rise housing and villa developments in European countries. No close-up detail of the panels is illustrated. Refer to www.sipbuildltd.com and download construction detail sheets.

Construction of a four-bedroom villa over an in-situ concrete basement and garage area. SIP panels manufactured in the UK and delivered to site in Cyprus.

Panel storage on site–wide variety of panel widths designed to suit house layout plan

General view of project under construction

Task 1: Fixing of sole plate on existing foundation slab
Tollerences to within + or – 5mm in line and level

Task 2: Erection and temporary propping of vertical wall panels on prepared sole plate. Design may incorporate steel columns and beams. Erection of walls to an external and internal wall layout plan

Engineered I joist roof

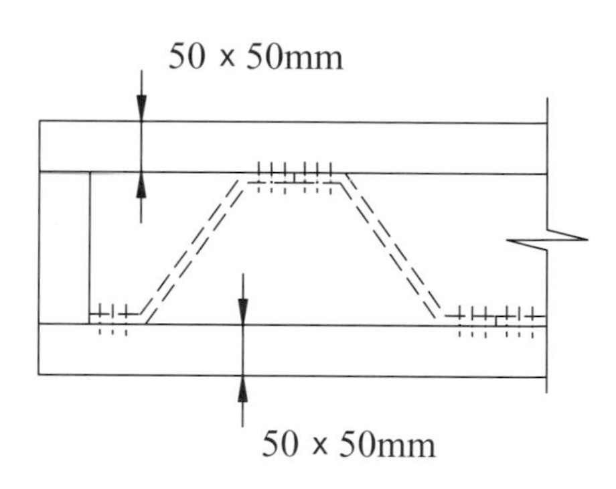

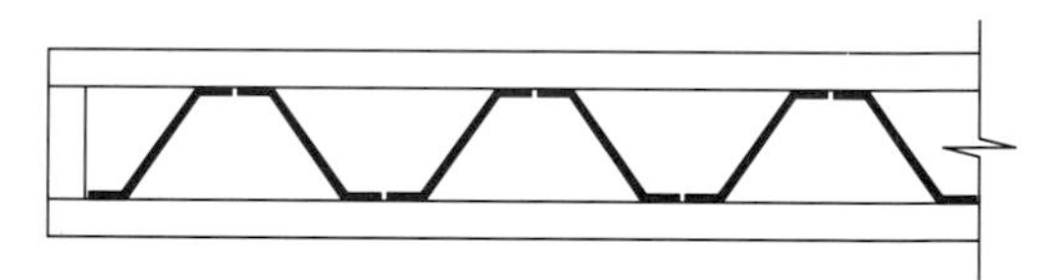

I joist beams at 450mm centres
Spans available up to 5 metes
Top surface covered in 20mm plywood, 100mm rigid insulation and roof finish

View of part completed villa

Many villa developments on sloping sites are constructed over in-situ basement areas. The erection of wall panels is shown at an advanced stage of construction and the roof construction is progressing. Engineering wood I joists are shown in position for the main roof spans and the projecting roof canopy areas.

CHAPTER THIRTEEN

DOMESTIC HOUSING CONSTRUCTION

13.0 Overview

The chapter tracks the construction of a pair of semi-detached houses from foundation level to the first fixing stage.

The images are divided into nine construction sequences with related construction sketches integrated into each section.

A schedule of current Building Regulations Parts A to P has been indicated. The application of the Regulations relative to Parts A, B, C, and E in relation to foundations and external wall and roof construction has been outlined.

A construction sequence diagram is illustrated together with a programme of work for the first 19 weeks of the project.

Section 9 of the construction stages covers work sequences to internal areas, i.e. first fixing works. It is important for students to understand the sequence of undertaking this stage of the works.

Additional sections on domestic construction have been included on domestic stair construction and forms of roof construction.

The construction practice sheets have been included to complete the domestic section.

13.1 Erection of a pair of semi-detached houses

Description of project

The project involves the construction of a pair of traditionally built houses for sale on completion of the building works.

Work is being undertaken by a small building firm consisting of two working tradesmen (i.e. bricklayers). Similar developments of up to six dwelling units have previously been successfully completed.

The houses have been designed by a local architectural service practice. Selling price per unit is to be in the order of £200,000, subject to market conditions.

Construction aspects

The houses are of traditional construction, i.e. externally finished in artificial stone with blockwork internal walls and concrete tiled roof.

The ground floor consists of a precast beam and block floor on traditional strip foundations.

Living accommodation is provided on the ground floor and first floor. The roof space provides a further bedroom and bathroom facility.

An external parking space is provided at the rear of the property contained within a small garden area.

Work stages

The progress of site works has been recorded in the following construction stages:

Sequence 1 – Foundations and ground floor
Sequence 2 – External walls - GF to first floor
Sequence 3 – First-floor construction
Sequence 4 – External walls - first floor to second floor
Sequence 5 – Second-floor construction
Sequence 6 – Gable wall construction
Sequence 7 – Roof construction
Sequence 8 – Roof tiling
Sequence 9 – Internal finishing work

The requirements of the current Building Regulations will be referenced to the appropriate stages of work.

13.2 Application of the Building Regulations

The application of the Building Regulations to the erection of a domestic dwelling is related to the various construction sequences.

Particular reference will be made to the following parts of the 2009 Regulations.

Part A –Structure

Part B –Fire safety

Part C –Site preparation and resistance to moisture

Part E –Sound insulation

Part L –Conservation of fuel and energy

Part M –Disabled access to and use of building

Site-recorded images will illustrate how the requirements of the Building Regulations are being addressed.

Domestic housing
Meeting the Building Regulations

Schedule of current Building Regulations 2009

Part A –Structure
Part B –Fire safety
Part C –Site preparation and resistance to moisture
Part D –Toxic substances
Part E –Sound insulation
Part F –Ventilation
Part G –Hygiene
Part H –Drainage and waste disposal
Part J –Combustion appliances and fuel storage
Part K –Protection from falling, collision and impact
Part L –Conservation of fuel and power
Part M –Disabled access to and use of building
Part N –Glazing
Part P –Electrical safety
Reg 7 –Materials and workmanship

Application of the Building Regulations

Thermal insulation (U value regulations)

Part L – 2009 Building Regulations

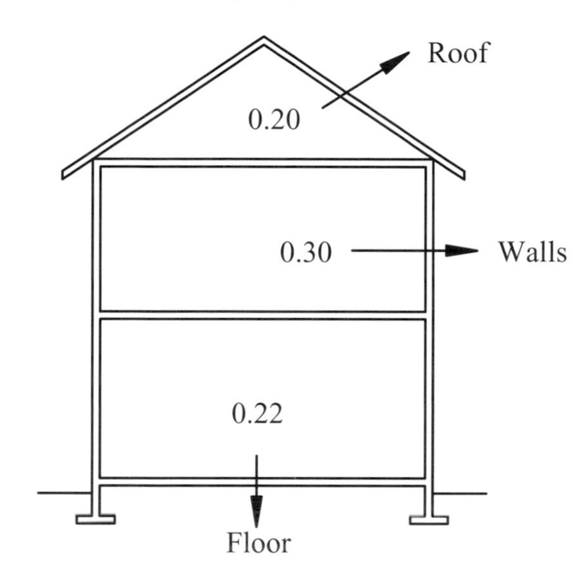

Fire safety (Part B)

External walls –1/2 hour
Party walls –1/2 hour
Roof –1 hour

Floors– fire separation between floors only applies to units of multi-occupancey, i.e. flats.

A fire stop is required between terraced or semi-detached houses, which is located along the party wall position and continued up the roof.

Sound requirements

Part E – 2009 Building Regulations

Semi-detached houses incorporate a party wall between dwellings. The wall should be constructed of 7N blocks and contain a 50mm clear cavity. The cavity should be constructed with special wall ties to reduce the passage of sound and requires to be cement rendered on both surfaces.

The cavity wall should be sound tested to record the resistance of airbourne sound.

Floors should be tested for both airbourne and impact sound by UKAS accredited body and results submitted to Building Control within 5 days of the test.

Ref. www.hush.uk.com/UK building regulation

Air testing

Part L 1A – Consevation of fuel and energy

The air testing of complete dwellings is a requirement of the building regulations. Building control will not provide a completion certificate without the necessary air testing results and report.

Tests cost in the order of £400 to £500 per dwelling type.

Domestic floor construction

Floor construction is required to meet the requirements of the Building Regulations with respect to:

Part A –Structural stability

Part B –Fire requirements

Part C –Moisture and dampness

Part E –Sound insulation–both impact and air bourne

This applies to all floor types, i.e. timber joist floors, in-situ and precast concrete floors.

For domestic dwellings 1/2 hour and 1 hour fire requirements may apply (which is relatively easy to achieve with both a timber and concrete floor).

A timber joist floor spanning 4m will be deemed to satisfy the building regulations when constructed is as shown.

Typical upper floor

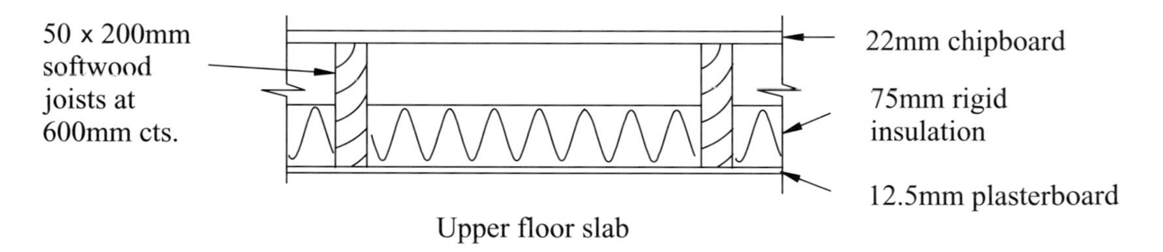

Upper floor slab

The construction illustrated meets Part A, B and E of the regulations for domestic dwellings.

For a solid ground floor slab, the following would be deemed to satisfy Part A, B, C and E.

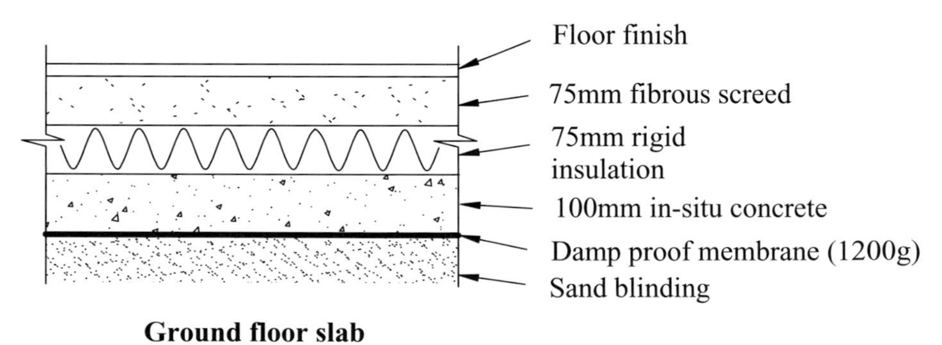

Ground floor slab

PART M – Disabled access to and use of building

Certain sections of part M apply to domestic housing with respect to disabled access. These include:

- Ramp to be provided to main building access of not greater than 1 in 12. Level access over the threshold to be provided.
- Door width to incorporate wheelchair access. Minimum clear width varies between 750–900mm depending on the angle of wheelchair approach to the door.
- All internal door widths to be a minimum of 750mm clear opening.
- Light switches to be positioned at a height of between 450–1200mm above floor level.
- Space for turning a wheelchair in dining and living room areas require a turning circle of 1500mm diameter.
- The living room should be located at the main entrance level.
- Specific wheelchair access requirments relate to ground floor toilets and bathrooms.

Ref: Building Regulations Part M
Lifetime house standards (recommendations)
On the Joseph Rowntrcc Foundation web site www.jrf.org.uk

13.3 Construction sequence and programme

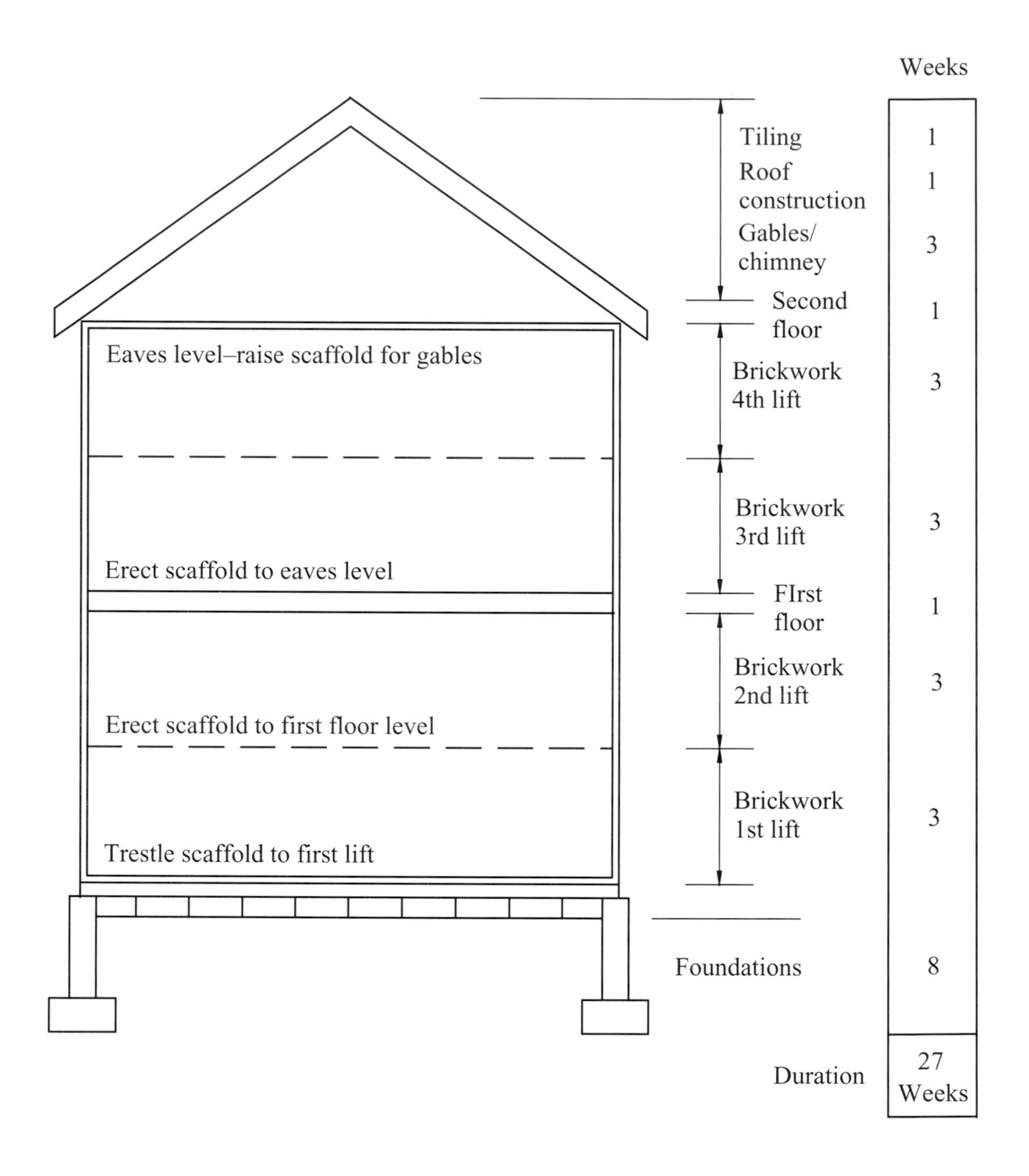

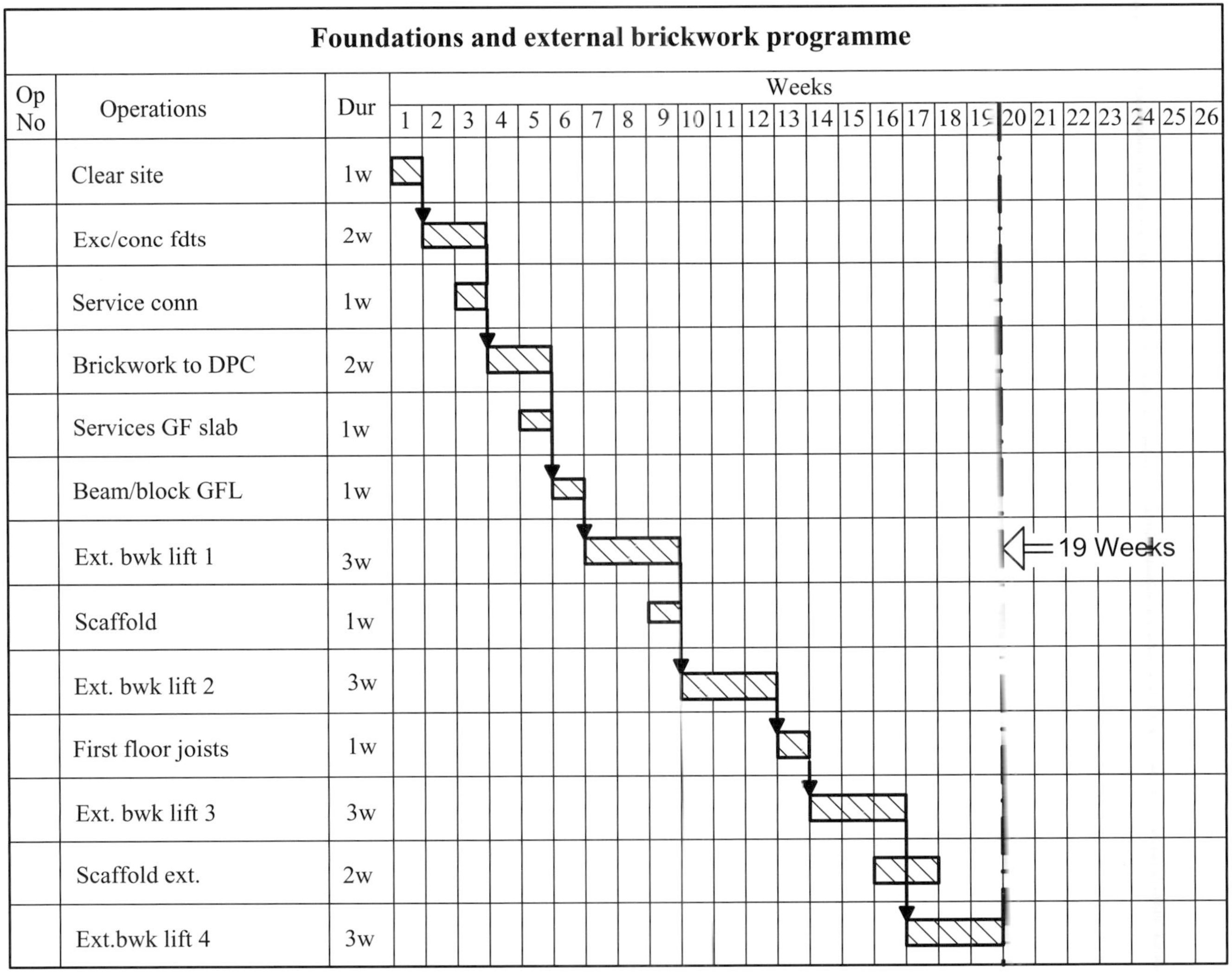

Foundations and external brickwork programme

Op No	Operations	Dur	Weeks 1	2	3	4	5	6	7	8	9	10	11	12	13	14	15	16	17	18	19	20	21	22	23	24	25	26
	Clear site	1w																										
	Exc/conc fdts	2w																										
	Service conn	1w																										
	Brickwork to DPC	2w																										
	Services GF slab	1w																										
	Beam/block GFL	1w																										
	Ext. bwk lift 1	3w																										
	Scaffold	1w																										
	Ext. bwk lift 2	3w																										
	First floor joists	1w																										
	Ext. bwk lift 3	3w																										
	Scaffold ext.	2w																										
	Ext.bwk lift 4	3w																										

19 Weeks

13.4 Sequence 1–foundations and ground floor

General view of ground floor slab.
Loading of blockwork in progress

Precast beam and block floor
Floor laid on traditional foundations

View of beam and block ground floor

Telescopic air vent at rear of air brick at DPC level (see construction detail)

Ground floor construction

Work up to the ground floor slab level was completed in 4 weeks period.
A section through the foundations and suspended beam and block floor is illustrated overleaf. A 1200g PVC radon barrier is shown in position together with the stepped DPC cavity tray. The radon barrier is to continue over the ground floor area incorporating taped joints.

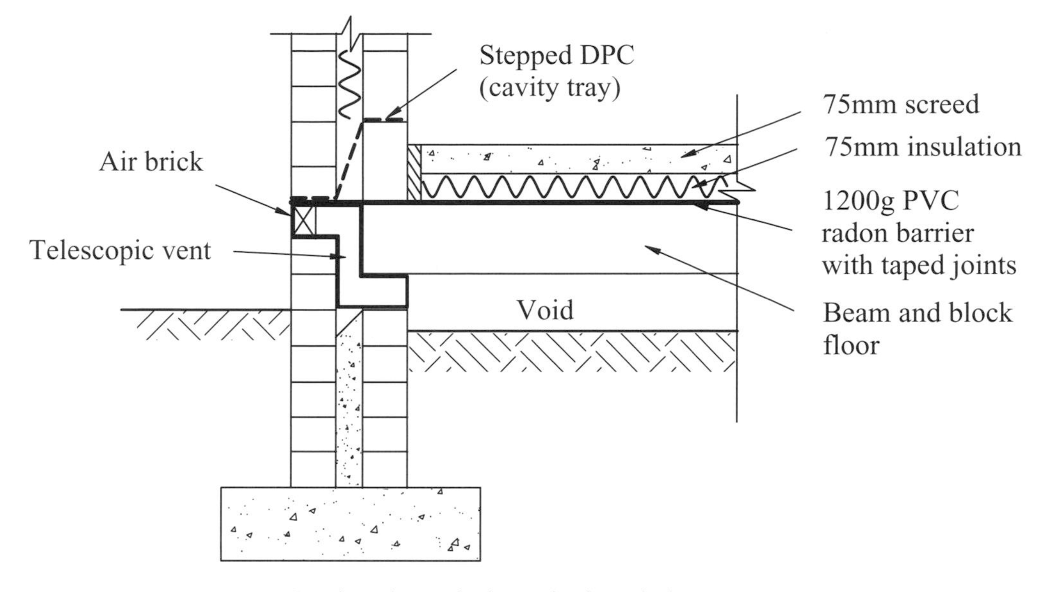

Section through the strip foundations
Telescopic vent shown in position

1200 gauge PVC radon barrier

Cavity tray formed at DPC level

Floor construction is required to meet the requirements of the Building Regulations with respect to:

Part A –Structural stability (either ground or suspended slab)

Part B –Fire requirements (1/2 hr fire resistance)

Part C –Moisture and dampness (provision of damp course membrane)

Part E –Sound insulation - both impact and airborne

This applies to all floor types, i.e. timber joist floors, in-situ and precast concrete floors and steel deck floors.

For domestic dwellings 1/2 hour and 1 hour fire requirements apply (which is relatively easy to achieve with both a timber and concrete floor).

Beam and block ground floor suspended slab

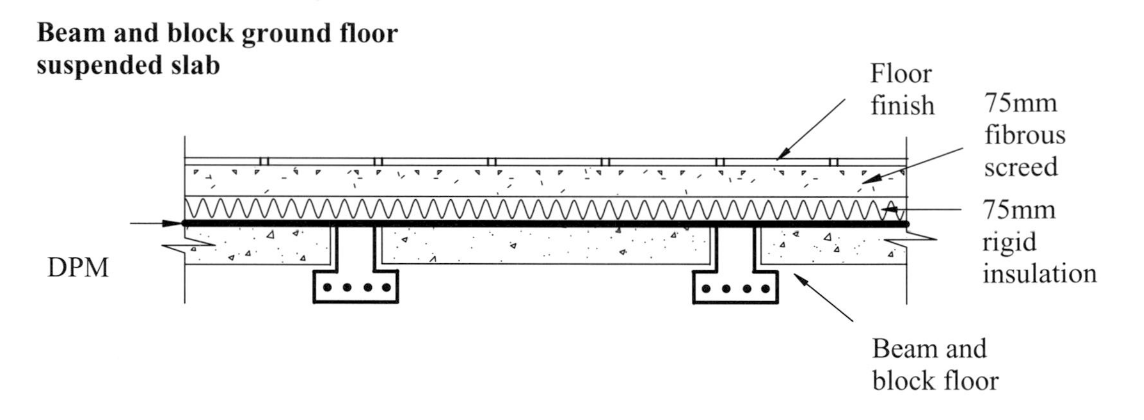

Alternative in-situ concrete slab Ground bearing slab

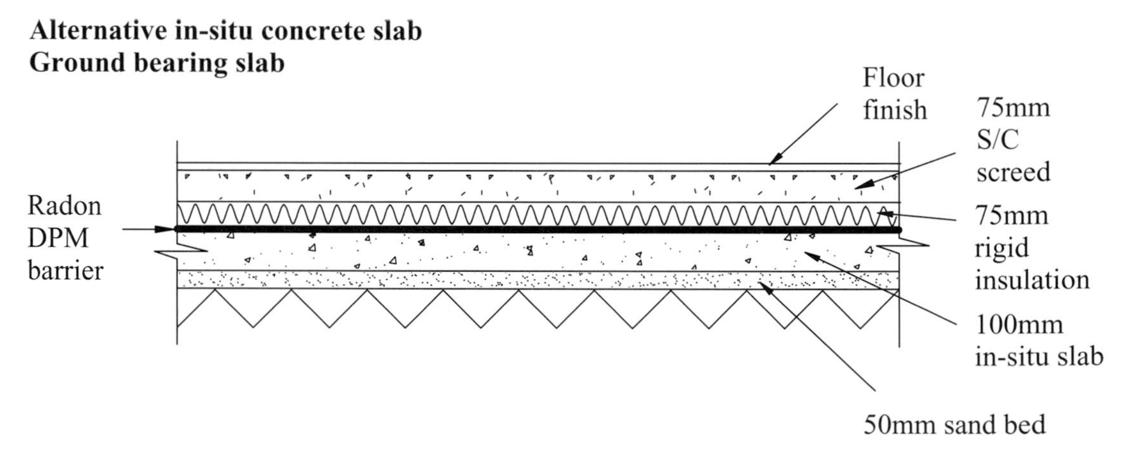

Both construction details meet Part A, B, C and E of the Building Regulations.

13.5 Sequence 2–external wall construction

Commencement of external walls above ground floor level

Plumbing corner of internal wall.
First lift of external wall blockwork

Return of blockwork at door opening to allow sealing of cavity

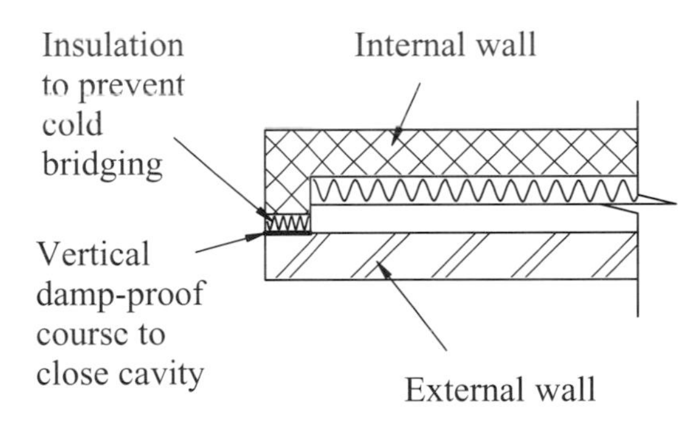

Door and window reveals

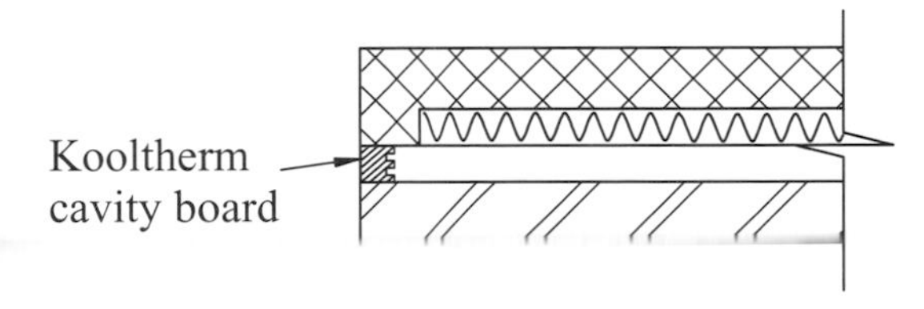

Insulated strip cavity closure

An alternative approach to closing cavity and reveals

Closing of cavity at window reveal with insulated DPC strip

Lintel over window openings at second floor level–cavity tray to be positioned over lintel

External wall construction

Alternative approach to closing cavity at reveal

Closing the cavity at the window and door reveal.

A wide range of cavity closers are available from specialist suppliers. These include trade names such as Cavalok and Thermabate etc.

These provide a rigid PVC template for the opening, damp proof course and thermal break at the reveal. They simplify the fixing of standard PVC windows. Ref. to web site index.

Use of PVC cavity closer (Kingspan/Thermabate)

Lintel over windows, firestop between dwellings

Galvanised lintel–bearing at each end

Section through insulated lintel

Steel galanised lintel over window opening.

Note: firestop (TDI cavity sock/barrier) in cavity at party wall position.

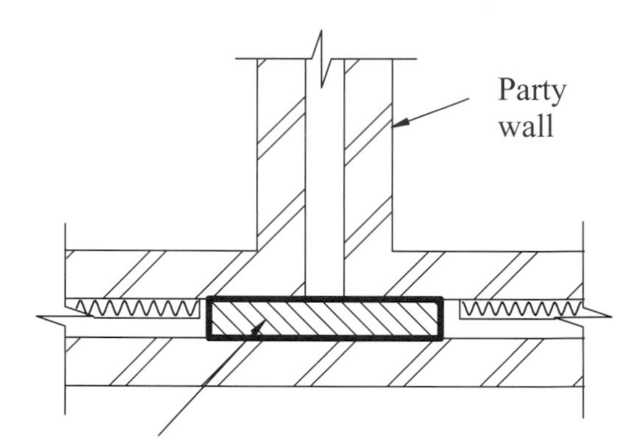

Firestop continued across party wall at roof level

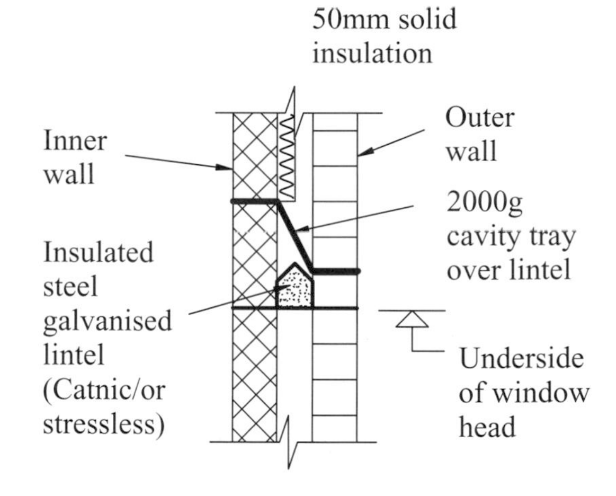

Lintel detail over window head

Construction detail at lintel position over windows

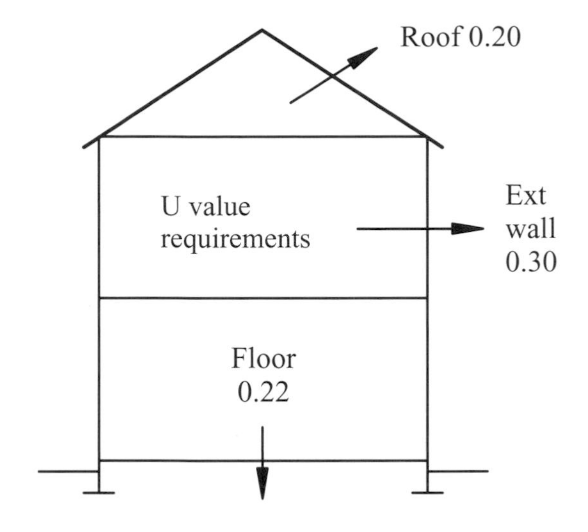

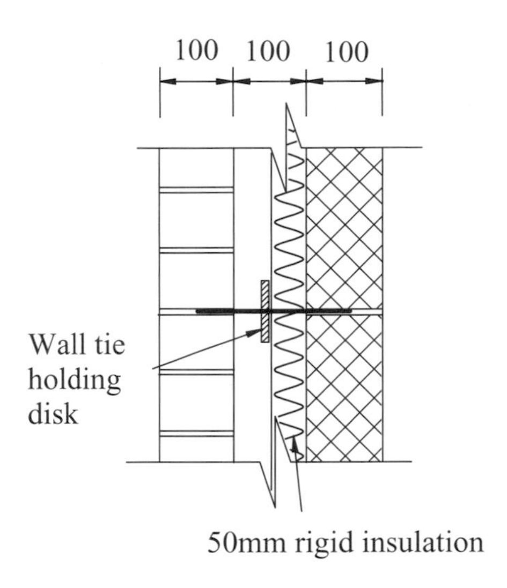

External
walls
U value
0.30
(alternative
solutions)

50mm rigid insulation to inner cavity wall, 50mm air space

Alternative–100mm full cavity insulation using Dritherm insulation bats

Meeting thermal insulation requirements

13.6 Sequence 3–First floor construction

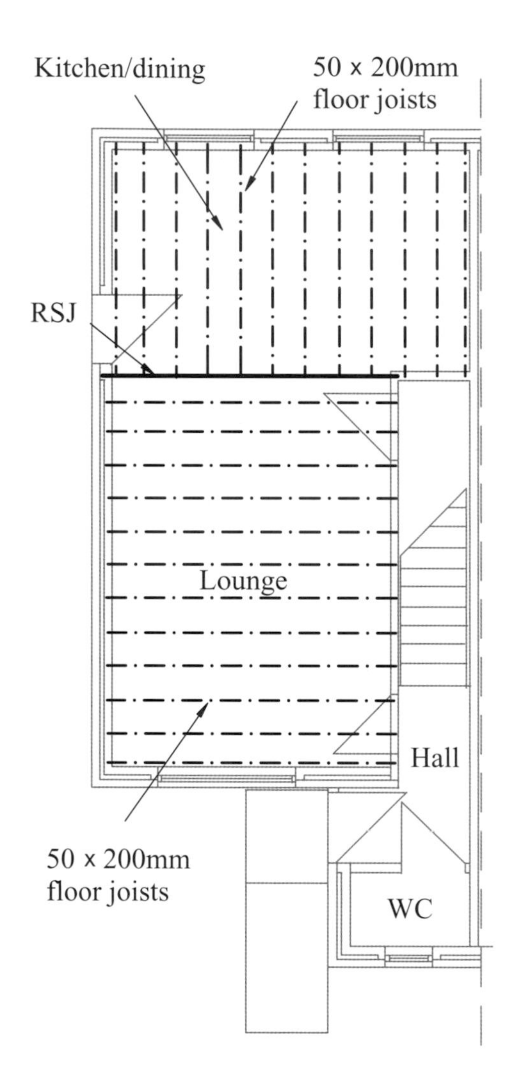

View of timber infill to steel beam to receive joist hangers

Close-up of joist hanger fixing

Plan of first floor

50 x 200mm softwood timber joists at 450mm centres

First floor layout

Arrangement of steel beams to support floor joists

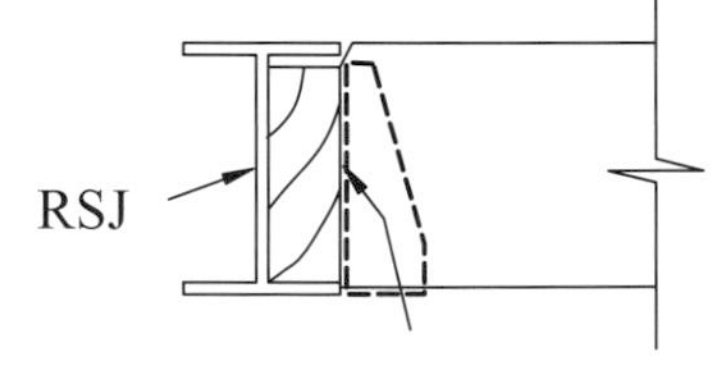

Steel beam on padstones with infill timber beam
50 x 200mm floor joists on joist hangers

First floor construction in 50 x 200mm timber joists

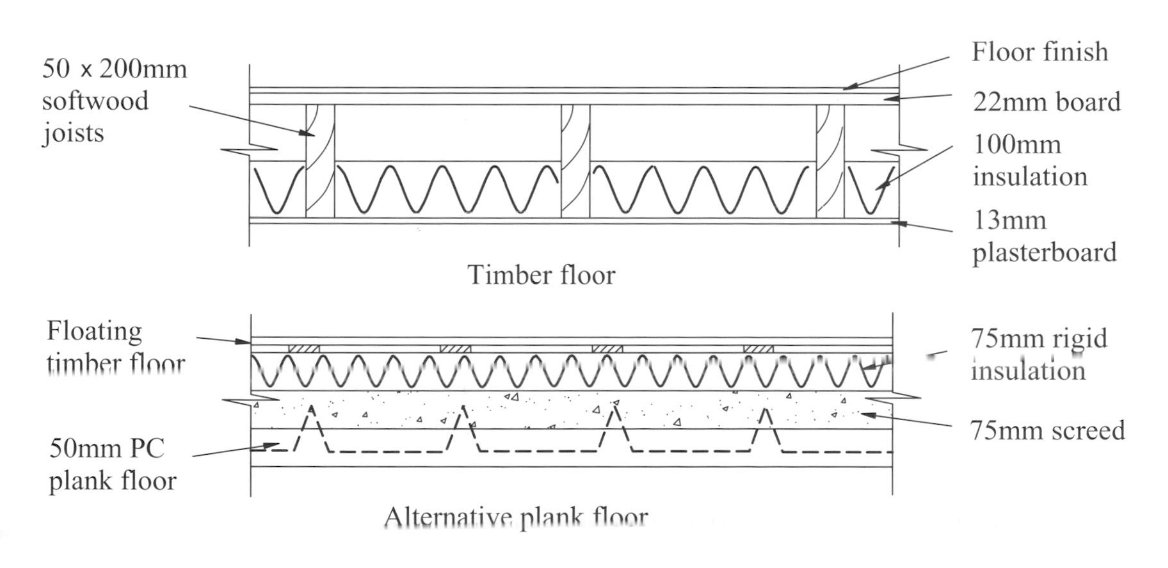

Deemed to satisfy fire and sound requirements (impact and airborne)

First floor construction

General view of softwood floor consisting of 50× 200mm timber joists at 450mm centres

First-floor joists being doubled up to accommodate load from internal partition

13.7 Sequence 4–external walls: first floor to second floor

Erection of blockwork to internal walls and stone facing externally. First floor to second floor erected in two lifts with bricklayers working off trestles for the internal wall blockwork.

Erect steelwork to support timber second floor. Layout of floor as the first-floor plan.

13.8 Sequence 5–second floor construction

Positioning and spacing of second floor joists
Joists spaced out by using standard blocks
Solid timber strutting between joists can be seen

Temporary timber gable frame erected to act as a guide for blockwork to gable wall. Timber trimming to staircase can be seen and the commencement of the gable blockwork.

13.9 Sequence 6–gable wall construction

Positioning of steel purlins spanning between gables
Wallplates bolted to top of steel flange of steel purlins

Party gable walls– gable chimney supported on additional steelwork.

13.10 Sequence 7–roof construction

General view of steel RSJ purlins with 100 x 75mm timber wallplates bolted to top of joist.

The 50 x 125mm rafters are notched and nailed to wallplates at 450mm centres.

Image shows the timber wallplate at eaves level together with the wall insulation chamfered back to accommodate the rafter ends.

Two carpenters completed the roof construction in a period of 8 days.

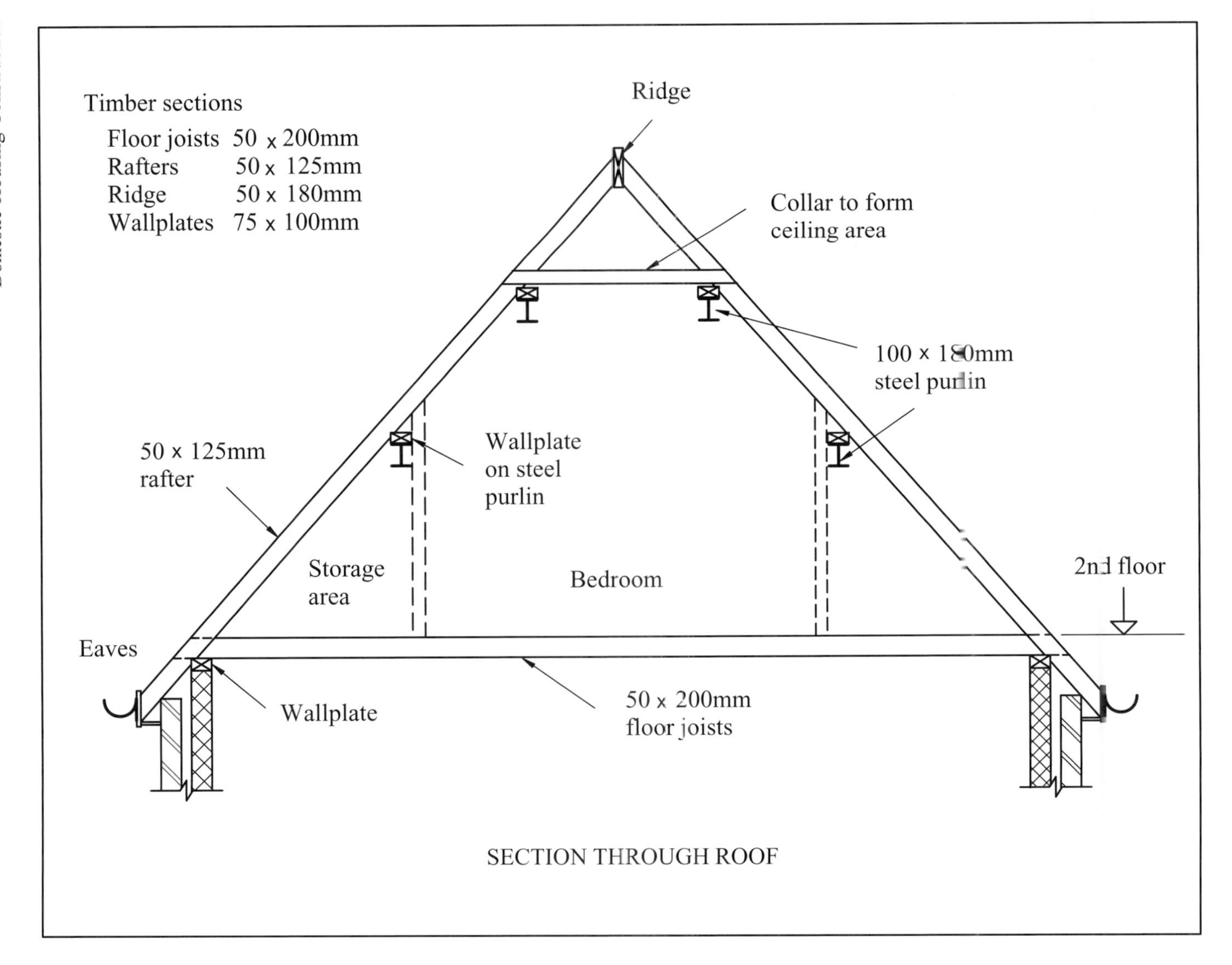
Timber sections
Floor joists 50 x 200mm
Rafters 50 x 125mm
Ridge 50 x 180mm
Wallplates 75 x 100mm
Ridge
Collar to form ceiling area
steel purlin
50 x 125mm rafter
Wallplate on steel purlin
Storage area
Bedroom
2nd floor
Eaves
Wallplate
50 x 200mm floor joists
SECTION THROUGH ROOF

Main roof construction showing 50 x 125mm rafters at 400mm centres notched to wallplates
Rafters located either side of party wall and at gable
Roofshield geotex sheathing layer and tile battens to be fixed as the following operation

Roof construction

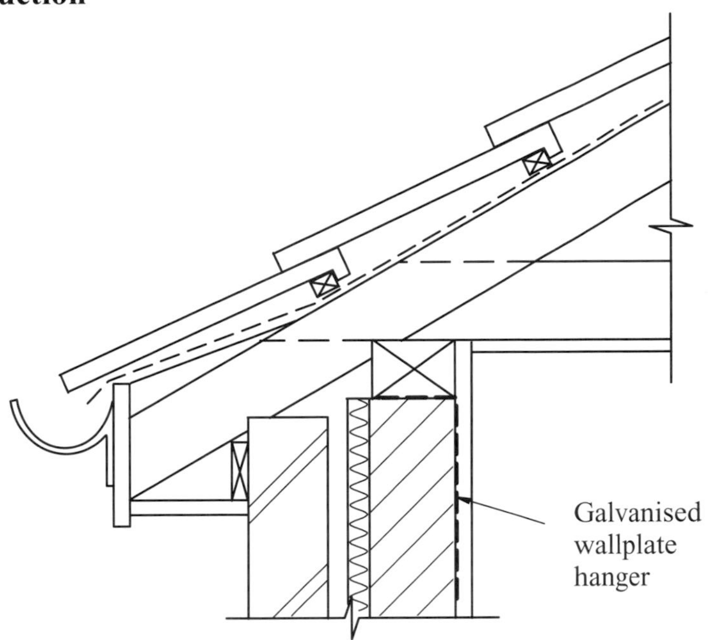

Eaves detail

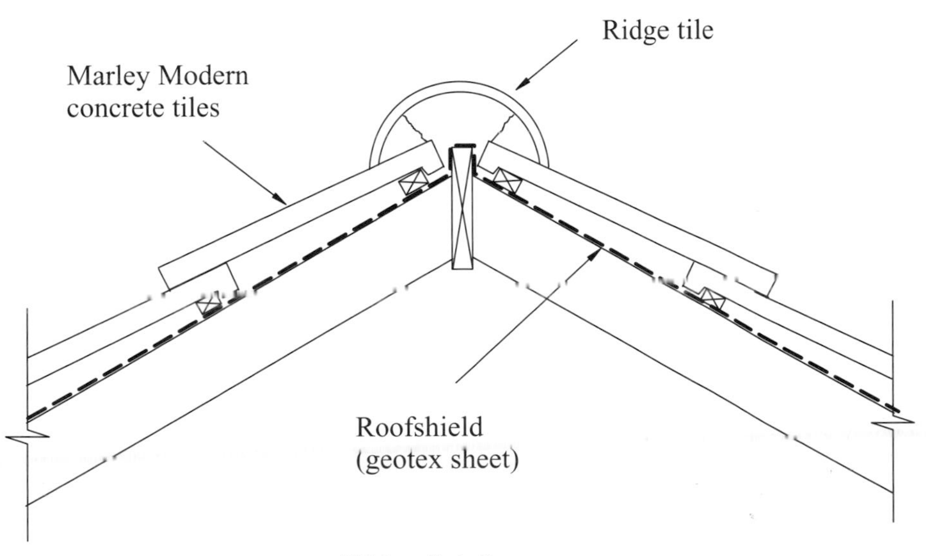

Ridge detail

13.11 Section 8–roof tiling

Boxed plastic eaves to receive 100mm gutter

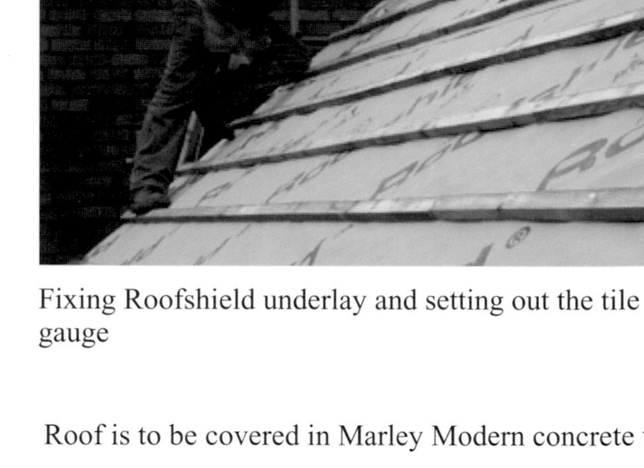

Fixing Roofshield underlay and setting out the tile gauge

Positioning Marley Modern concrete tiles

Roof is to be covered in Marley Modern concrete tiles.

The roof was felted (Roofshield underlay) battened and tiled in 3 days (two tilers and one labourer).

Tiles have interlocking edge and top nib for fixing over tile batten

Part completed roof tiling

13.12 Sequence 9–internal finishing work

Once the roof tiling has been completed and the building is watertight, work to the internal areas may commence.

Four stages of work are normally involved in completing interior fitting works.

Stage 1–First fix work
Stage 2–Internal plastering to walls and ceilings
Stage 3–Second fix work
Stage 4–Final fix work

These stages cannot be clearly defined and separated as a certain degree of overlap will occur due to each development being somewhat different.

Stage 1–first fix work

- Fitting external doors and windows
- Fitting internal staircase access
- Work to internal partitions (block or stud)
- First fix electrics and plumbing
- Plasterboard to ceilings and partitions
- Connecting mains services
- First fix joiner (door linings and window boards)

Stage 2–internal plastering to walls and ceilings

- Plasterboard and skimming partitions may be undertaken by plasters

Stage 3–second fix work

- Second fix electrical and plumbing
- Second fix carpenter/joiner
- Second fix bathroom and toilet areas
- Second fix kitchen (kitchen units and worktops)
- Floor tiling

Stage 4–final fix work

- Kitchen fittings and specialist equipment
- Fitting door ironmongery
- Internal wardrobes and fit-out

Internal finishing work

Fitting external windows and doors

Fitting internal stairs for access purposes

Boarding to floors

Fixing internal stud wall partitions

Insulating sloping ceiling area

Internal finishing work

Fixing extract ducts in ceiling space

Mounting electrical sockets on internal partitions

Fixing insulation to ceiling and wall areas

Electrical wiring to plug sockets in partitions

Plastic service pipe runs to radiator connections, sinks etc

Electric service runs between floor joists

Internal finishing work

75mm insulation bats to partitions and 100mm Dritherm loose insulation to ceiling and floor areas. Service connections to bathroom waste outlets can be seen.

Insulation to ceiling/floor areas

Plasterboard complete to upstairs room areas

13.13 Domestic stair construction

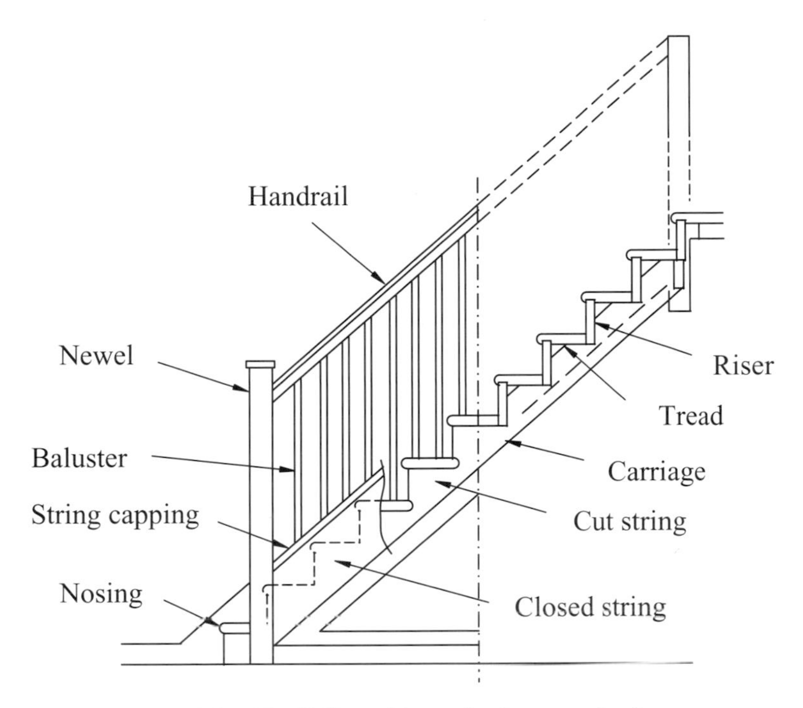

The Building Regulations relative to staircase construction

Building Regulations

- Staircases should have a maximum rise of 220mm and a minimum going of 220mm.
- They should have a maximum pitch of 42°.
- Flights should have a handrail on at least one side if they are less than 1m wide and on both sides if they are wider than 1m.
- Handrails on stairs and landings should have a minimum height of 900mm.
- No openings of any balustrading should allow the passage of a 100mm sphere.
- A minimum of 2000mm of clear headroom is required above the pitch line.

  For further details, see Regulations and Approved Document K.

Staircase layout

A wide variety of staircase layouts are available including:

- straight flight
- half landing flight
- quarter landing
- quarter landing with winders
- straight flight to quarter landing then flights going left and right, i.e. Y shaped

Staircase types

Winder flight

Y shaped staircase

Open string stair

Straight flight

Spiral staircase springing from steel column

Steel support to hardwood treads

13.14 Forms of roof construction

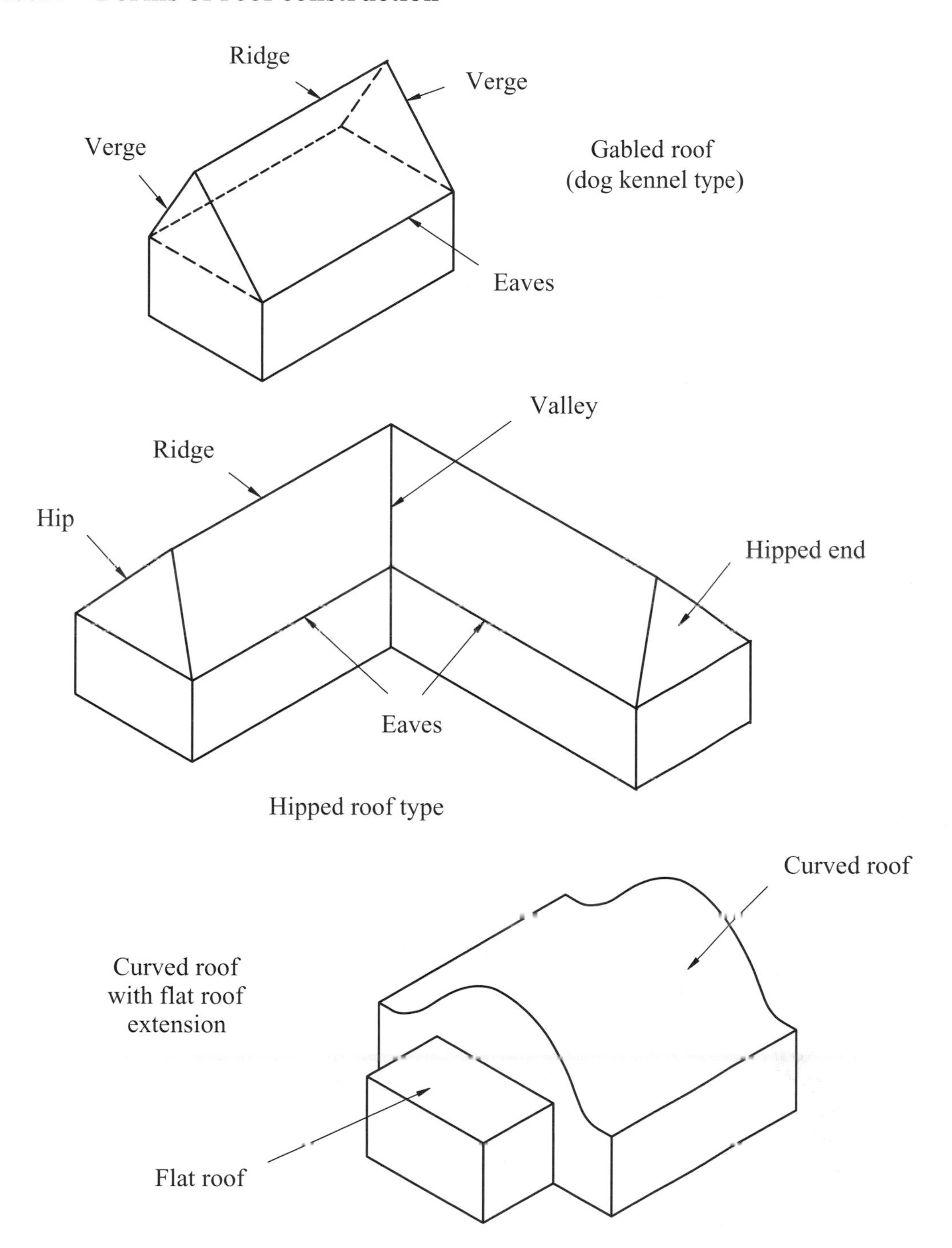

13.15 Gabled roof construction

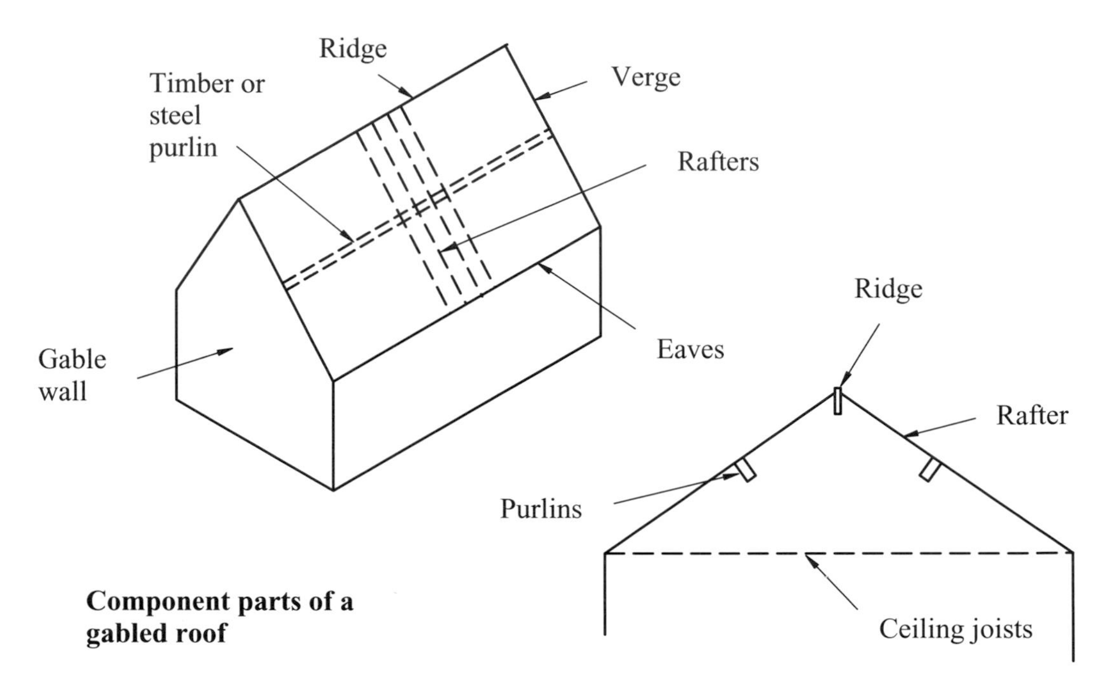

Component parts of a gabled roof

Gabled roof under construction 225 × 75mm rafters with RSJ steel forming ridge

Insulation to gabled roof

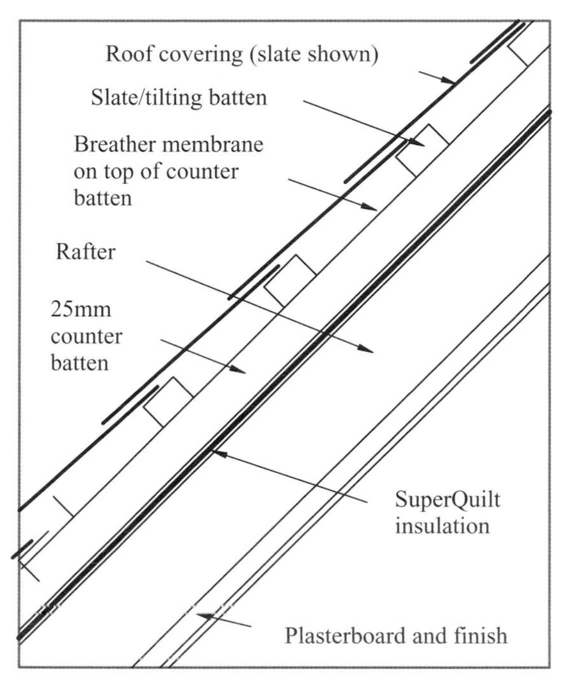

SuperQuilt over rafter application

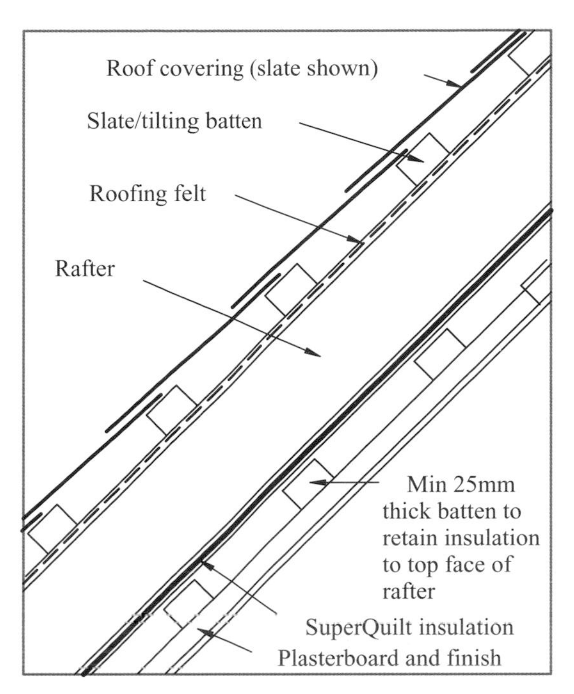

SuperQuilt under rafter application

Fixing SuperQuilt over rafter–counter battens shown in position

Roof openings to accommodate Velux roof lights
Galvanised joist hangers fitted to support trimmers to openings

Roof finished in reclaimed stone roofing slabs

Stone slabs drilled to accommodate aluminium toggle bolt from which slabs are positioned on roof battens, i.e. hooked over roof batten and held in position by overlapping slab

Velux roof lights

Four Velux roof lights have been incorporated in the rear roof slope. This allows ventilation and light into the room area below. Details of the secret gutter and PVC coated metal flashing around the Velux are shown.

13.16 Formation of a manhole base and connections

Stage 1

Stage 2

Stage 3

Stage 1

Extend main drain into position at manhole. Note use of trench boxes and manhole boxes to ensure safe working conditions.

Stage 2

Place dry concrete mix to form base of manhole and locate main half-round channel and connection, i.e. first 225mm diameter pipe.

Stage 3

Position all other connections, i.e. three 100mm diameter connections shown.

Stage 4

Place additional dry concrete mix around connections. Shape to form benching in bottom of manhole prior to placing precast manhole ring.

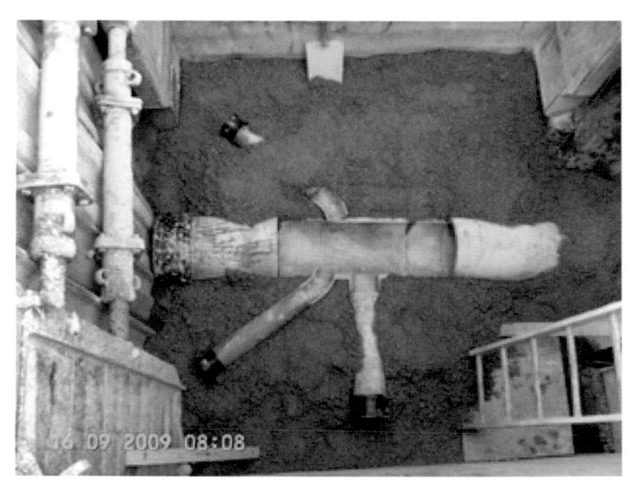

Stage 4

Stage 5

Stage 5

Place precast concrete manhole ring in position using crane.

Complete benching up to base of manhole and finish with smooth sand/cement mix.

Stage 6

Place manhole cover slab in position.

Stage 7 and 8

Place rubber sheet around manhole perimeter (to act as formwork) and partially backfill to hold in position.

Place concrete in surround to manhole after part backfilling to support rubber shutter.

Stage 6

Stage 7

Stage 8

13.17 Insulating a room

Case study

Insulating an existing garage to provide habitable accommodation for a playroom and office area.

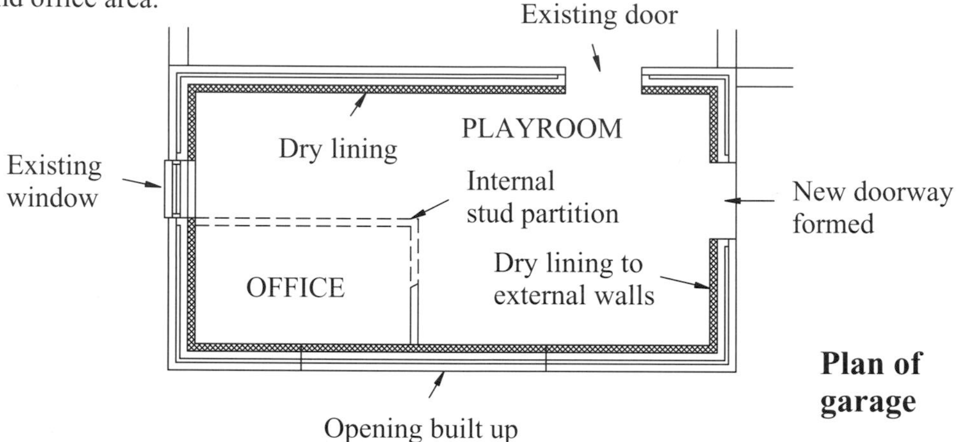

Plan of garage

Rigid insulation

Quinn Thermal Rigid PIR insulation board is specified for the wall lining. The insulation board is 2.4m x 1.2m in size and 70mm in thickness. It is delivered to site, shrink wrapped for protection from the weather. Cost per sheet approximately £40.

Blanket insulation

Superglass fibreglass insulation 150mm thick is positioned between the existing ceiling joists. The ceiling is to be lined internally with 13mm plasterboard complete with a plaster skim coat.

Fixing stud partitions

The 50 x 100mm softwood timber studs are fixed around the internal perimeter of the area at 450mm centres. A timber sole and a head piece are incorporated together with a mid row of studs.

The 70mm-thick rigid insulation board is cut and positioned between the studs to enclose the room.

Internal view of stud partition forming inner office area.

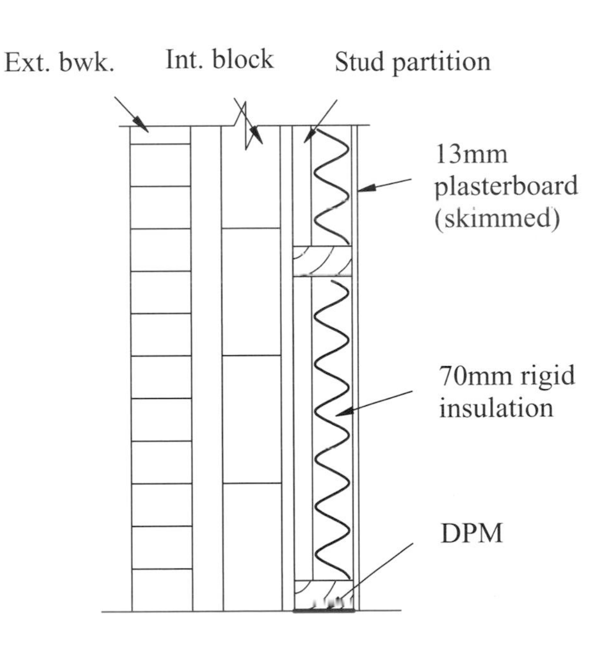

Section through external wall and stud partition wall

Publications by the author and others

1979 - Cost and Financial Control for Construction Firms
B. Cooke and W.B. Jepson
Macmillan Press
ISBN 0-333-24096-0

1981 - Contract Planning and Contractual Procedures
B. Cooke
Macmillan Press
ISBN 0-333-30720-8

1988 - Contract Planning Case Studies
B. Cooke
Macmillan Press
ISBN 0-333-44547-3

1994 - European Construction
B. Cooke and G. Walker
Macmillan Press
ISBN 0-333-59465-7

1998 - Construction Planning, Programing and Control
B. Cooke and P. Williams
Macmillan Press
ISBN 0-333-67736-7

2009 - Construction Planning, Programing and Control, Third Edition
B. Cooke and P. Williams
Wiley-Blackwell
ISBN 1-4051-8380-2

WEB SITE REFERENCES – IN TOPIC AREAS

This section will prove of interest to students preparing coursework's and projects on construction related areas.

Comments have been included to outline specific areas of interest – and notes on the usefulness of the web site.

BUILDING FRAMES

www.corusconstruction.com/architectural (Teaching resource guide)
www.corusconstruction.com/longspanstructures
www.dawsonera.com (Architectural designs in steel)
www.deltastructures.com (American Space Frame company)
www.framing@metsec.com (Steel framed modular buildings)
www.glulam.co.uk (Glulam laminated structures)
www.greatbuildings.com (Sainsbury Centre – a space framed roof structure)
www.info@timber-frame (Flight Timber Structures)
www.ketchum.org (Shell roof types)
www.lilleheden.dk (Danish timber laminated 3 pin arches)
www.mero,com (Space frame website)
www.meroframes.com (Analysis and design of space frames)

FLOOR CONSTRUCTION

www.bison.co.uk (Covers hollow floors / hollow composite and solid composite floors)
www.longley.uk.com (Illustrates four types of floor construction including polystyrene formers)
www.thomasarmstrong.co.uk (An excellent student reference covering a wide range of beam and block floors, wide slab floors and pre-cast concrete wall systems)

FORMWORK AND SCAFFOLDING

www.doka.com (Access products on Doka International – full range of floor, wall and climbing formwork systems indicated. In the Doka climbing systems extensive construction detail illustrated)
www.info@peri.de
www.mabey.co.uk (Formwork supplier / hire company)
www.pcharrington.com (Slipform moving formwork system for lift shafts)
www.peri.de (German formwork systems – CD available)

GENERAL TECHNOLOGY

www.Wikipedia.org/wik/sheffield_winter_gardens

A construction encyclopaedia on all aspects of construction – try accessing great buildings or recently completed building projects – like, The Sheffield Winter Gardens project. – a wealth of project data and photographs during construction becomes available. Now try The Sainsbury Building

INSULATION OF BUILDINGS

www.kingspan.com (A full range of products and solutions for roof systems)
www.cavalok.com (Ideal site for a range of cavity closers with case study applications)
www.cellotex.co.uk (Cellotex board products)
www.energysavers.gov/your_home/insulation (An excellent summary of insulation applications to domestic dwellings – five types of insulation outlined)
One can access all the Kingspan product range. Try accessing details of Kingspan Top Deck 1000 product (An insulated roof deck system)
www.kingspan.com/topdeckKS1000/insulateddecking
www.planetinsulation.co.uk (Series of articles on insulating your home using Gyproc, Kingspan and Cellotex products. Table of pay-back periods shown for various methods of providing sources of energy and heat source considerations)
www.renovation_headquarters.com/insulation/types (Types of insulation part 1-indicates the wide range of products available)
www.rockwool.co.uk (Fibreglass bats and quilts)
www.superquilt.co.uk (Indicating the range of Superquilt products – the reflective silver paper quilt)

PILING AND FOUNDATIONS

www.bascol.co.uk (Bachy Soletanche are large specialist piling contractors offering a wide range of techniques including bored pile walls and diaphragm wall construction)
www.keller-ge.co.uk (Keller ground engineering indicates a wide range of case studies)
www.roger-bullivant.co.uk (Excellent UK site for piling, foundation packages and mini piling techniques – range of case studies available)
www.mkpiling.com (A North West based piling firm offering precast/CFA piling/mini-pile systems)
www.projectpiling.co.uk (A fast moving display of a range of projects in North West England. Interesting range of site photographs of foundation work including the use of permanent formwork solutions)
www.westpile.co.uk (An overview of the main piling techniques with diagrams and related contract case studies. Insert /index.php/projects/The_Rose_bowl_building_Leeds and view the piling process on site. An overview of various piling systems is reviewed by a series of case studies)

PRE-CAST BASEMENTS

www.apc-concrete.co.uk (Part of the Thomas Armstrong Group -Pre-cast concrete design and construct provider – Case study included in the text based on the underground storage tank))
www.basements.org (The Basement Information Centre – an excellent range of case studies available on UK projects showing detailed sequences of construction)
www.glatthaar.co.uk (A leading German manufacturer with UK outlets – good range of applications to domestic construction)

ROOFING SYSTEMS

www.kingspan.com (Full range of Kingspan roofing products and solutions. Insulation requirements for floors, walls and ceilings)
www.ribaproductselector.com/trocal (Trocal is a single membrane system for industrial roofing applications (Try using www.ribaproductselector.com to trace details of building products and systems – simply enter the product name.)
www.sarnafil.com (A single layer high quality thermoplastic membrane for flat roof construction) – Part of the SIKA products group.
www.ward-insulated-panels (Ward top deck roofing system, full range of roofing systems with case studies)

SITE ACCOMMODATION

www.allspace.ie
www.avflex.co.uk (Case studies available on the web site)
www.portacabin.co.uk (UK's leading manufacturer of mobile accommodation)
www.ukcabins.com (Mobile site accommodation)
www.wacouk.com

TIMBER FRAMED CONSTRUCTION

www.borderoak.co.uk (Traditional oak framed cottages/ houses and structures – heavy oak timber framing based on the post and beam method – excellent and informative site)
www.flighttimber.com (Interesting range of case studies. SIPS Eco panels described)
www.huf-haus.co.uk (The famous German Post and Beam Framed House). Try accessing The Self Builders Guide/4 Homes/Channel 4 on the web page)
www.mcveighinsulation.co.uk (Major SIP panel fabricator for office, school projects and commercial buildings – Excellent informative web site with extensive case studies – North West based company)
www.taylor-lane.co.uk (Good photo library and case studies included/ video clip from live project)
wwww.trada.co.uk (Good range of current case studies and text book references)

WASTE MANAGEMENT

www.aggregain-agency.gov.uk (Step by step tool for dealing with construction waste)
www.ciwm.co.uk (The web site of the Chartered Institute of Waste Management. Enter waste explained option for comprehensive overview of duty of care/hazardous waste)
www.eauc.org.uk (Defra guidance on site waste management plans)
www.netregs.gov.uk (Waste management licensing regulations)
www.premierwaste.co.uk (An excellent web site with case studies of waste management services provided in the North West region – Case Study included in text on waste transfer station)

ADDITIONAL WEB REFERENCES

www.o'sullivancivils.co.uk Wide range of basement, piling and earthworks projects illustrated. Excellent basement formwork applications. Sixteen case studies available to access.
www.sipbuildltd.com Informative site with a wide range of applications. Standard building detail shown for frames and external finishes. Full technical data available on fire, sound and air tightness requirements.

Students preparing construction projects should find these web sites of interest.

GLOBAL COMMUNICATION

Theories, Stakeholders, and Trends

THOMAS L. McPHAIL

University of Missouri at St. Louis

ALLYN AND BACON

Boston ▪ London ▪ Toronto ▪ Sydney ▪ Tokyo ▪ Singapore

Series Editor: *Molly Taylor*
Editorial Assistant: *Michael Kish*
Editorial-Production Service: *Omegatype Typography, Inc.*
Composition and Prepress Buyer: *Linda Cox*
Manufacturing Buyer: *Julie McNeill*
Cover Administrator: *Kristina Mose-Libon*
Electronic Composition: *Omegatype Typography, Inc.*

A Pearson Education Company
75 Arlington Street
Boston, MA 02116

Internet: www.ablongman.com

Between the time Website information is gathered and then published, it is not unusual for some sites to have closed. Also, the transcription of URLs can result in unintended typographical errors. The publisher would appreciate notification where these occur so that they may be corrected in subsequent editions. Thank you.

Library of Congress Cataloging-in-Publication Data

McPhail, Thomas L.
Global communication: theories, stakeholders, and trends/Thomas L. McPhail.
p. cm.
Includes bibliographical references and index.
ISBN 0-205-15635-5
1. Communication, International. I. Title.
P96.I5 M37 2002
302.2–dc21

2001022428

Printed in the United States of America
10 9 8 7 6 5 4 3 2 06 05 04